W9-CCN-389

Absolutely Small

Absolutely Small

How Quantum Theory Explains
Our Everyday World

Michael D. Fayer

ΛMΛCOM

American Management Association

New York • Atlanta • Brussels • Chicago • Mexico City • San Francisco
Shanghai • Tokyo • Toronto • Washington, D.C.

Bulk discounts available. For details visit:
www.amacombooks.org/go/specialsales
Or contact special sales:
Phone: 800-250-5308
E-mail: specialsls@amanet.org
View all the AMACOM titles at: www.amacombooks.org

This publication is designed to provide accurate and authoritative information in regard to the subject matter covered. It is sold with the understanding that the publisher is not engaged in rendering legal, accounting, or other professional service. If legal advice or other expert assistance is required, the services of a competent professional person should be sought.

Library of Congress Cataloging-in-Publication Data

Fayer, Michael D.
  Absolutely small : how quantum theory explains our everyday world / Michael D. Fayer.
    p. cm.
  Includes bibliographical references and index.
  ISBN-13: 978-0-8144-1488-0
  ISBN-10: 0-8144-1488-5
  1. Quantum theory—Popular works. 2. Quantum chemistry—Popular works.
I. Title.
QC174.12.F379 2010
530.12—dc22

                                                    2009044707

About AMA
American Management Association (www.amanet.org) is a world leader in talent development, advancing the skills of individuals to drive business success. Our mission is to support the goals of individuals and organizations through a complete range of products and services, including classroom and virtual seminars, webcasts, webinars, podcasts, conferences, corporate and government solutions, business books, and research. AMA's approach to improving performance combines experiential learning—learning through doing—with opportunities for ongoing professional growth at every step of one's career journey.

Printing number

10 9 8 7 6 5 4 3 2

Contents

Preface

IF YOU ARE READING THIS, you probably fall into one of two broad categories of people. You may be one of my colleagues who is steeped in the mysteries of quantum theory and wants to see how someone writes a serious book on quantum theory with no math. Or, you may be one of the vast majority of people who look at the world around them without a clear view of why many things in everyday life are the way they are. These many things are not insignificant aspects of our environment that might be overlooked. Rather, they are important features of the world that are never explicated because they are seemingly beyond comprehension. What gives materials color, why does copper wire conduct electricity but glass doesn't, what is a trans fat anyway, and why is carbon dioxide a greenhouse gas while oxygen and nitrogen aren't? This lack of a picture of how things work arises from a seemingly insurmountable barrier to understanding. Usually that barrier is mathematics. To answer the questions posed above—and many more—requires an understanding of quantum theory, but it actually doesn't require mathematics.

This book will develop your quantum mechanics intuition, which will fundamentally change the way you view the world. You have an intuition for mechanics, but the mechanics you know is what we refer to as classical mechanics. When someone hits a long drive baseball, you know it goes up for a while, then the path turns over and the ball falls back to Earth. You know if the ball is hit harder, it takes off faster and will go farther before it hits the ground. Why does the ball behave this way? Because gravity is pulling it back to Earth. When you see the moon, you know it is orbiting the Earth. Why? Because gravity attracts the moon to the Earth. You don't sit down and start solving Newton's equations to calculate what is going on. You know from everyday experience that apples fall down not up and that if your car is going faster it will take longer to stop. However, you don't know from everyday experience why cherries are red and blueberries are blue. Color is intrinsically dependent on the quantum mechanical description of molecules. Everyday experience does not prepare you to understand the nature of things around you that depend on quantum phenomena. As mentioned here and detailed in the book, understanding features of everyday life, such as color or electricity, requires a quantum theory view of nature

Why no math? Imagine if this book contained discussions of a topic that started in English, jumped into Latin, then turned back to English. Then imagine that this jumping happened every time the details of an explanation were introduced. The language jumping is what occurs in books on quantum theory, except that instead of jumping from English to Latin, it jumps from English to math. In a hard core quantum mechanics book (for example, my own text, *Elements of Quantum Mechanics* [Oxford University Press, 2001]), you will find things like, "the interactions are described by the following set of coupled differential equations." After the equations, the text reads, "the solutions are," and more equations appear. In contrast, the presentation in this book is descriptive. Diagrams replace the many equations, with the exception of some small algebraic equations—and these simple equations are explained in detail.

Even without the usual overflow of math, the fundamental philosophical and conceptual basis for and applications of quantum theory are thoroughly developed. Therefore, anyone can come away with an understanding of quantum theory and a deeper understanding of the world around us. If you know a good deal of math, this book is still appropriate. You will acquire the conceptual understanding necessary to move on to a mathematical presentation of quantum theory. If you are willing to do some mental gymnastics, but no math, this book will provide you with the fundamentals of quantum theory, with applications to atomic and molecular matter.

Absolutely Small

1

Schrödinger's Cat

WHY ARE CHERRIES RED and blueberries blue? What is the meaning of size? These two questions seem to be totally unrelated. But, in fact, the second question doesn't seem to be a question at all. Don't we all know the meaning of size? Some things are big, and some things are small. But, the development of quantum theory showed that the first two questions are intimately related and that we had a completely false concept of size until a couple of decades into the twentieth century. Our ideas about size, if we thought about size at all, worked just fine in our everyday lives. But beginning in approximately 1900, the physics that was used to describe all of nature, and the physics that still works remarkably well for landing a spacecraft on Mars, began to fall apart. In the end, a fundamentally new understanding of size was required not only to explain why cherries are red and blueberries are blue, but also to understand the molecules that make up our bodies, the microelectronics that run our computers, why carbon dioxide is a greenhouse gas, and how electricity can move through metals. Our everyday experiences teach us to think in terms of classical physics, the physics that was greatly advanced and formalized by Sir Isaac Newton (1642–1727). Everything we know from early childhood prepares us to view nature in a manner that is fundamentally wrong. This book is about the concept of absolute size and its consequence, quantum theory, which requires us to fundamentally change our way of thinking about nature. The first half of the book describes the basic concepts of quantum theory. The second half applies quantum theory to many aspects of the world around us through an examination of the properties of atoms and molecules and their roles in everyday life.

This book began with a simple question. Does quantum mechanics make sense? I was asked to address this question at "Wonderfest 2005, the Bay Area Festival of Science," sponsored by the University of California at Berkeley Department of Physics and the Stanford University Department of Chemistry. Wonderfest is a yearly event that presents a variety of lectures on "the latest findings" in a number of fields to an audience of nonspecialists. However, I was not asked to discuss the latest findings in my own research, but the topic, "does quantum mechanics make sense," which has been argued about by scientists and laypeople alike since the inception of quantum theory in 1900. In addition, I had only one-half hour to present my affirmative answer to the question. This was a tall order, so I spent several months thinking about the subject and a great deal of time preparing the lecture. After the event, I thought I had failed—not because it is impossible to make plain the important issues for nonspecialists, but because the time constraint was so severe. To get to the crux of the matter, certain concepts must be introduced so that contrasts between classical mechanics and quantum mechanics can be drawn. This book is my opportunity to address the quantum theory description of nature with sufficient time to do the subject justice. The book uses very simple math involving at most small equations. The idea is to make quantum theory completely accessible to the nonscientist. However, the fact that the book requires essentially no math does not mean that the material is simple. Reading Kierkegaard requires no math but is not simple. However, unlike Kierkegaard, the meaning of the material presented below should be evident to the reader who is willing to do a little mental exercise.

Classical mechanics describes the motion of a baseball, the spinning of a top, and the flight of an airplane. Quantum mechanics describes the motion of electrons and the shapes of molecules such as trans fats, as well as electrical conductivity and superconductivity. Classical mechanics is a limiting case of quantum mechanics. Quantum mechanics contains classical mechanics but not vice versa. In that respect, classical mechanics is wrong. However, we use classical mechanics to design bridges, cars, airplanes, and dams. We never worry about the fact that the designs were not done using the more general description of nature embodied in quantum theory. The use of classical mechanics will not cause the bridges to collapse, the cars to crash, the airplanes to fall from the sky, or the dams to burst. In its own realm, the realm of mechanics that we encounter in everyday life, classical mechanics works perfectly. Our intuitive feel of how the world works is built up from everyday experiences, and those experiences are, by and large, classical. Nonetheless, even in everyday life classical mechanics cannot explain why the molecules in a blueberry make it blue and the molecules in a cherry make it red. The instincts we have built up over a lifetime of observing certain aspects of nature leave us unprepared to intuitively understand other aspects of nature, even though such aspects of nature also pervade everyday life.

SCHRÖDINGER'S CAT

Schrödinger's Cat is frequently used to illustrate the paradoxes that seem to permeate the quantum mechanical description of nature. Erwin Schrödinger (1887–1961) and Paul A.M. Dirac (1902–1984) shared the Nobel Prize in Physics in 1933 for their contributions to the development of quantum theory, specifically "for the discovery of new productive forms of atomic theory." Schrödinger never liked the fundamental interpretation of the mathematics that underpins quantum theory. The ideas that bothered Schrödinger are the exact topics that will be discussed in this book. He used what has come to be known as "Schrödinger's Cat" to illustrate some of the issues that troubled him. Here, Schrödinger's Cat will be reprised in a modified version that provides a simple illustration of the fact that quantum mechanics doesn't seem to make sense when discussed in terms of everyday life. The cats offered here are to drive the issues home and are not in Schrödinger's original form, which was more esoteric. The scenario presented will be returned to later. It will be discussed as an analogy to real experiments explained by quantum theory, but not as an actual physical example of quantum mechanics in action.

Imagine that you are presented with 1000 boxes and that you are going to participate in an experiment by opening them all. You are told that there is a half-dead cat in each box. Thus, if you opened one of the boxes, you might expect to find a very sick cat. Actually, the statement needs to be clarified. The correct statement is that each of the cats is not half dead, but rather each cat is in a state that is simultaneously completely dead and perfectly healthy. It is a 50-50 mixture of dead and healthy. In other words, there is a 50% chance that it is dead and a 50% chance that it is alive. Each of the thousand cats in the thousand boxes is in the exact same state. The quantum experimentalist who prepared the boxes did not place 500 dead cats in 500 boxes and 500 live cats in the other 500 boxes. Rather, he placed identical cats that are in some sense 50-50 mixtures of dead and perfectly healthy in each box. While the cats are in the closed boxes, they do not change; they remain in the live-dead mixed state. Furthermore, you are told that when you open a box and look in, you will determine the cat's fate. The act of looking to see if the cat is alive will determine if the cat is dead or alive.

You open the first box, and you find a perfectly healthy cat. You open the next three boxes and find three dead cats. You open another box and find a live cat. When you are finished opening the 1000 boxes, you have found 500 live cats and 500 dead cats. Perhaps, more astonishing, would be if you start again with a new set of one 1000 boxes, each containing again a 50-50 mixture of live-dead cats. If you open the boxes in the same order as in the first trial, you will not necessarily get the same result for any one box. Say box 10 in the first run produced a live cat on inspection. In the second run, you may find it produced a dead cat. The first experimental run gives you no information on what any one box will contain the second time. However, after opening all 1000 boxes on the second run, you again find 500 live cats and 500 dead cats.

I have to admit to simplifying a little bit here. In two runs of the Schrödinger's Cats experiment, you probably would not get exactly 500 live and 500 dead cats on each run. This is somewhat like flipping an honest coin 1000 times. Because the probability of getting heads is one half and the probability of getting tails is one half, after 1000 flips you will get approximately 500 heads. However, you might also get 496 heads or 512 heads. The probability of getting exactly 500 heads or 500 live cats out of 1000 trials is 0.025 (2.5%). The probability of getting 496 heads is 0.024 (2.4%) and 512 heads is 0.019 (1.9%). The probability of getting only 400 heads or 400 live cats out of 1000 trials is $4.6 \times 10^{-11} = 0.000000000046$. So the probable outcomes are clustered around 500 out of a 1000 or 50%. Knowing that you have 1000 Schrödinger's Cat boxes with 50-50 mixtures of live-dead cats or 1000 flips of an honest coin, you can't say what will happen when you open one box or flip the coin one time. In fact, you can't even say exactly what will happen when you open all 1000 boxes or flip the coin 1000 times. You can say what the probability of getting a particular result is for one event and what the likely cumulative results will be for many events.

NOT LIKE FLIPPING COINS

A fundamental difference exists between Schrödinger's Cats, or more correctly real quantum experiments, and flipping pennies. Before I flip a penny, it is either heads or tails. When I flip it, I may not know what the outcome will be, but the penny starts in a well-defined state, either heads or tails, and ends in a well-defined state, either heads or tails. It is possible to construct a machine that flips a penny so precisely that it always lands with the same result. Nothing inherent in nature prevents the construction of such a machine. If a penny with heads up is inserted into the machine, a switch could determine whether the penny lands heads or tails. In flipping a coin by hand, the nonreproducibility of the flip is what randomizes the outcome. However, a box containing Schrödinger's Cat is completely different. The cat is a 50-50 mixture of live and dead. It is the act of opening the box and observing the state of the cat that causes it to change from a "mixed state" into a "pure state" of either alive or dead. It doesn't matter how precisely the boxes are opened. Unlike flipping pennies, a machine constructed to open each of the 1000 boxes exactly the same way will not make the results come out the same. The only thing that can be known about opening any one box is that there is a 50% chance of finding a live cat.

REAL PHENOMENA CAN BEHAVE LIKE SCHRÖDINGER'S CATS

As described, the Schrödinger's Cat problem cannot be actualized. However, in nature many particles and situations do behave in a manner analogous to opening Schrödinger's Cat boxes. Particles such as photons (particles of light), electrons, atoms, and molecules have "mixed states" that become "pure states" upon observation, in a manner like that described for Schrödinger's Cats. The things that make up everyday matter, processes, and phenomena behave at a fundamental level in a way that, at first, is as counterintuitive as Schrödinger's Cats. However, the problem does not lie with the behavior of electrons and atoms, but rather with our intuition of how things should behave. Our intuition is based on our everyday experiences. We take in information with our senses, which are only capable of observing phenomena that involve the behavior of matter governed by the laws of classical mechanics. It is necessary to develop a new understanding of nature and a new intuition to understand and accept the quantum mechanical world that is all around us but not intuitively understandable from our sensory perceptions.

2

Size Is Absolute

THE FUNDAMENTAL NATURE OF SIZE is central to understanding the differences between the aspects of the everyday world that fit into our intuitive view of nature and the world of quantum phenomena, which is also manifested all around us. We have a good feel for the motion of baseballs, but we mainly gloss over our lack of knowledge of what gives things different colors or why the heating element in an electric stove gets hot and glows red. The motion of baseballs can be described with classical mechanics, but color and electrical heating are quantum phenomena. The differences between classical and quantum phenomena depend on the definition of size.

The quantum mechanical concept of size is the correct view, and it is different from our familiar notion of size. Our common concept of size is central to classical mechanics. The failure to treat size properly, and all of the associated consequences of that failure, is ultimately responsible for the inability of classical mechanics to properly describe and explain the behavior of the basic constituents of matter. A quantum mechanical description of matter is at the heart of technological fields as diverse as microelectronics and the computer design of pharmaceuticals.

SIZE IS RELATIVE IN EVERYDAY LIFE

In classical mechanics, size is relative. In quantum mechanics, size is absolute. What does relative versus absolute size mean, and why does it matter?

In classical mechanics and in everyday life, we determine whether something is big or small by comparing it to something else. Figure 2.1 shows two rocks. Looking at them, we would say that the rock on the left is bigger than the rock on the right. However, because there is nothing else to compare them to, we can't tell if they are what we might commonly call a big rock and a small rock. Figure 2.2 shows the rock on the left again, but this time there is something to compare it to. The size of the rock is clear because we have the size of a human hand as a reference. Because we know how big a typical hand is, we get a good feel for how big the rock is relative to the hand. Once we have the something against which to

FIGURE 2.1. *Two rocks.*

make a size comparison, we can say that the rock is relatively small, but not tiny. If I were to describe the rock over the phone, I could say it is somewhat bigger than the palm of your hand, and the person I am talking to would have a good idea of how big the rock is. In the absence of something of known size for comparison, there is no way to make a size determination.

Figure 2.1 demonstrates how much we rely on comparing one thing to another to determine size. In Figure 2.1, the two rocks are on a white background, with no other features for reference. Their proximity immediately leads us to compare them and to decide that the rock on the left is larger than the rock on the right. Figure 2.3 shows the rock on the right in its natural setting. Now we can see that it is actually a very large rock. The hand on the rock gives a very good reference from which to judge its size. Like the rock in the hand, the rock with the hand on top provides us with a scale that permits a relative determination of size. It is clear from these simple illustrations that under normal circumstances, we take size to be relative. We know how big something is by comparing it to something else.

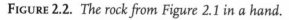

FIGURE 2.2. *The rock from Figure 2.1 in a hand.*

FIGURE 2.3. *The other rock from Figure 2.1, but now in a context from which its size can be judged.*

OBSERVATION METHOD CAN MATTER

Why does the definition of size, relative versus absolute, matter? To observe something, we must interact with it. This is true in both classical and quantum mechanics.

Figure 2.4 illustrates the observation of a rose. In a totally dark room, we cannot see the rose. In Figure 2.4, however, light emanating from the light bulb falls on the rose. Some of the light is absorbed, and some of it bounces off. (Which colors are absorbed, and therefore, which colors bounce off to make the leaves look green and the petals look red, is a strictly quantum mechanical phenomenon that will be discussed in Chapter 8.) A portion of the light that

FIGURE 2.4. *The light bulb illuminates the rose. The light that bounces off the rose enters the eye, enabling us to see the rose.*

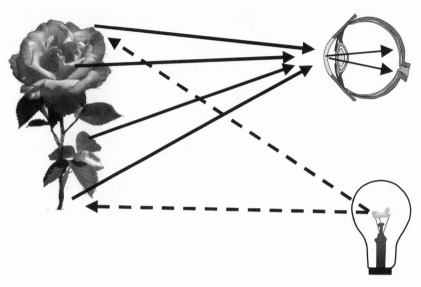

bounces off is detected by the eye and processed by the brain to observe the rose. The observer is interacting with the rose through the light that bounces off of it.

Once we recognize that we must interact with an object to observe it, we are in a position to define big and small. The definition of what is big and what is small is identical in classical mechanics and quantum mechanics. If the disturbance to an object caused by an observation, which is another way of saying a measurement, is negligible, then the object is big. If the disturbance is nonnegligible, the object is small. In classical mechanics, we make the following assumption.

Assume: When making an observation, it is always possible to find a way to make a negligible disturbance.

If you perform the correct experiment, then the disturbance that accompanies the measurement is negligible. Therefore, you can observe a system without changing it. However, if you do the wrong experiment in trying to study a system, you make a nonnegligible disturbance, and the object is small. A nonnegligible disturbance changes the system in some way and, it is desirable, if possible, to make a measurement that doesn't change the thing you are trying to measure. Classical theory assumes that you can reduce the size of the disturbance to be as small as desired. No matter what is under observation, it is possible to find an experimental method that will cause a negligible disturbance. This assumed ability to find an experiment that produces a negligible disturbance implies that *size is only relative*. The size of an object depends on the object and on your experimental technique. There is nothing inherent. Any object can be considered to be big by observing it with the correct method, a method that causes a negligible disturbance.

Suppose you decide to examine the wall of a room in which you are sitting by throwing many billiard balls at it. In your experiment, you will observe where the balls land after they bounce off of the wall. You start throwing balls and, pretty soon, plaster is flying all over the place. Holes appear in the wall, and the balls you throw later on don't seem to bounce off the same way the earlier balls did. This may not be surprising because of the gaping holes that your measurement method is making in the wall. You decide that this is not a very good experiment for observing the wall. You start over again after having a good painter restore the wall to its original state. This time you decide to shine light on the wall and observe the light that bounces off of the wall. You find that this method works very well. You can get a very detailed look at the wall. You observe the wall with light for an extended period of time, and the properties you observe do not change.

BIG OR SMALL—IT'S THE SIZE OF THE DISTURBANCE

When the wall was observed with billiard balls, it was small because the observation made a nonnegligible disturbance. When the wall was observed with light, it was big. The observation made a negligible disturbance. In these experiments, which can be well described with classical mechanics, the wall's size was relative. Do the poor experiment (observation with billiard balls), and the wall is small. Do a good experiment (observation with light), and the wall is big.

In classical mechanics, there is nothing intrinsic about size. Find the right experiment, and any object is big. It is up to the experimental scientist to design or develop the right experiment. Nothing intrinsic in classical mechanics theory prevents a good experiment from being performed. A good experiment is one that produces a negligible disturbance during the measurement. In other words, a good experiment does not change the object that is being observed, and, therefore, the observation is made on a big object.

CAUSALITY FOR BIG OBJECTS

The importance of being able to make any object big is that it can be observed without changing it. Observing an object without changing it is intimately related to the concept of causality in classical mechanics. Causality can be defined and applied in many ways. One statement of causality is that equal causes have equal effects. This implies that the characteristics of a system are caused by earlier events according to the laws of physics. In other words, if you know in complete detail the previous history of a system, you will know its current state and how it will progress. This idea of causality led Pierre-Simon, Marquis de Laplace (1749–1827), one of the most renowned physicists and mathematicians, to declare that if the current state of the world were known with complete precision, the state of the world could be computed for any time in the future. Of course, we cannot know the state of the world with total accuracy, but for many systems, classical mechanics permits a very accurate prediction of future events based on accurate knowledge of the current state of a system. The prediction of the trajectory of a shell in precision artillery and the prediction of solar eclipses are examples of how well causality in classical mechanics works.

As a simple but very important example, consider the trajectory of a free particle, such as a rock flying through space. A free particle is an object that has no forces acting on it, that is, no air resistance, no gravity, etc. Physicists love discussing free particles because they are the simplest of all systems. However, it is necessary to point out that a free particle never really exists in nature. Even a rock in intergalactic space has weak gravity influencing it, weak light shining on it, and occasionally bumps into a hydrogen atom out there among the galaxies. Nonetheless, a free particle is useful to discuss and can almost be realized in a laboratory. So our free particle is a hypothetical true free particle despite its impossibility.

The free particle was set in motion some time ago with a momentum p, and at the time we will call zero, t = 0, it is at location x. x is the particle's position along the horizontal axis. The trajectory of the rock is shown in Figure 2.5 beginning at t = 0. The momentum is p = mV where m is the mass of the object and V is its velocity. The mass on earth is just its normal weight. If the rock is on the moon, it has the same mass, but it would have one-sixth the weight because of the weaker pull of gravity on the moon.

A very qualitative way to think about momentum is that it is a measure of the force that an object could exert on another object if

FIGURE 2.5. *A free particle in the form of a rock is shown moving along a trajectory.*

they collided. Imagine that a small boy weighing 50 pounds runs into you going 20 miles per hour. He will probably knock you down. Now imagine that a 200-pound man runs into you going 5 miles per hour. He will probably also knock you down. The small boy is light and moving fast. The man is heavy and moving slow. Both have the same momentum, 1000 lb−miles/hour. (lb is the unit for pound.) In some sense, both would have the same impact when they collide with you. Of course, this example should not be taken too literally. The boy might hit you in the legs while the man would hit you in the chest. But in a situation where these types of differences did not occur, either would have essentially the same effect when running into you.

Momentum is a vector because the velocity is a vector. A vector has a magnitude and a direction. The velocity is the speed and the direction. Driving north at 60 mph is not the same as driving south at 60 mph. The speed is the same, but the direction is different. The momentum has a magnitude mV and a direction because the velocity has a direction. In Figure 2.5, the motion is from left to right across the page.

At $t = 0$ we observe (make measurements of) the rock's position and momentum. Once we know x and p at $t = 0$, we can predict the trajectory of the rock at all later times. For a free particle, predicting the trajectory is very simple. Because there are no forces acting on the particle, no air resistance to slow it down or gravity to pull it down to earth, the particle will continue in a straight line indefinitely. At some later time called t' (t prime), $t = t'$, the rock will have moved a distance $d = Vt'$. The distance is the velocity multiplied by how long the particle has been traveling. Since we started at time equal to zero, $t = 0$, then t' is how long the particle has been moving—for example, one second. So at time t' we know exactly where to look for the rock. We can make an observation to see if the particle is where we think it should be and, sure enough, there it is, as shown in Figure 2.5. We can predict where it will be at a later time, and observe that it is in fact there. This is shown on the right side of Figure 2.5. We have predicted where the particle will be, and when we make an observation, it is there. The rock is traveling with a well-defined trajectory, and the principle of causality is obeyed.

NONNEGLIGIBLE DISTURBANCES MATTER

Now consider Figure 2.6. The rock is prepared identically to the situation shown in Figure 2.5. At t = 0, it has position x and momentum p. Again it is observed at t = t'.

Its position is as predicted from the values of x and p at t = 0. However, some time after t = t', a bird flies into a rock. (You will have to forgive my drawing of the bird. This is the best I can do on a computer with a mouse.) In the jargon of physics, we might refer to this as a bird-rock scattering event. The bird hitting the rock makes a nonnegligible disturbance. Therefore, it is not surprising that a measurement of the position and momentum made some

FIGURE 2.6. *A free particle in the form of a rock is moving along a trajectory. At time t = 0, it has position x and momentum p. At a later time, t = t', it has moved to a new position where it is observed, and its future position is predicted. However, some time later, a bird flies into the rock. The prediction made at t' is no longer valid.*

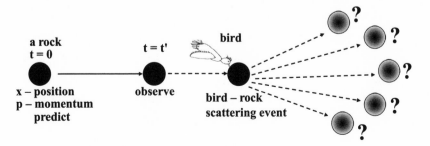

time after the scattering event will not coincide with the predictions made based on the trajectory determined at t = 0. According to the precepts of classical mechanics, if we knew everything about the bird, the rock, and how they interact (collide with each other), we could make a prediction of what would happen after the bird-rock scattering event. We could test our predictions by observation. Observation is possible in classical mechanics because we can find a method for observation that makes a negligible disturbance of the system. That is, we can always find a way to make the system big. But the important point is that following a nonnegligible disturbance, it is not surprising that predictions are not fulfilled, as they were based on the known trajectory that existed prior to the disturbance.

THERE IS ALWAYS A DISTURBANCE

Quantum theory is fundamentally different from classical mechanics in the way it treats size and experimental observation; the difference makes size absolute. Dirac succinctly put forward the assumption that makes size absolute.

> *Assume: There is a limit to the fineness of our powers of observation and the smallness of the accompanying disturbance, a limit that is inherent in the nature of things and can never be surpassed by improved technique or increased skill on the part of the observer.*

This statement is a wild departure from classical thinking. It says that whenever you observe a system (make a measurement), there is always a disturbance; it may be small, but it is always there. The size of this disturbance is part of nature. No improvements in instrumentation or new methods of observation can make this minimum disturbance vanish or become smaller.

SIZE IS ABSOLUTE

Dirac's statement has ramifications that are part of all formulations of quantum theory. His assumption immediately makes size absolute. An object is big in the absolute sense if the minimum disturbance that accompanies a measurement is negligible. An object is small in the absolute sense if the inherent minimum disturbance is not negligible. At the most fundamental level, classical mechanics is not set up to describe objects that are small in the absolute sense. In classical mechanics, any object can be made "big" by finding the right experiment to use in making an observation. In the development of classical mechanics, it was never envisioned that because of the inherent properties of nature, it was impossible to improve methodology to the point where an observation did not change a system. Therefore, classical mechanics is not set up to deal with objects that are small in an absolute sense. Its inability to treat objects that are absolutely small, such as electrons or atoms, is the reason that classical mechanics fails when it is applied to the description of such objects.

Figure 2.7 illustrates the nature of the problem. An electron is a particle that is small in the absolute sense. (Later we will discuss in detail the meaning of the word particle, which is not the same as the classical concept of particle.) At t = 0, it is moving along a trajectory. As with the rock, we want to see if it is actually doing what we think it is doing so that we can make subsequent predictions. We use the least invasive method to observe the electron; we let it interact with a single particle of light, a photon. (Below is a detailed discussion of the nature of light and what it means to have a particle of light.) Here is what makes this problem completely different from that illustrated in Figure 2.5. Because an electron is absolutely small, even observing it with a single particle of light causes a nonnegligible disturbance. The electron is changed by the observation. We cannot make subsequent predictions of what it will

FIGURE 2.7. *At time, t = 0, an electron is moving along some trajectory. At time, t = t', we observe it in a minimally invasive manner by letting it interact with a single particle of light, a photon. (Photons are discussed in detail later.) The electron-photon interaction causes a nonnegligible disturbance. It is not possible to make a causal prediction of what happens after the observation.*

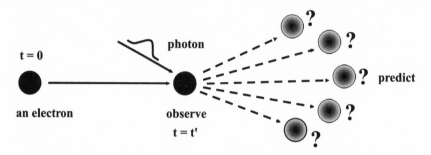

do once we observe it to see if it is doing what we think it is doing. Causality applies to undisturbed systems. The act of observing the electron disturbs it. You can predict what a system is doing as long as you don't look to see if it is actually doing what you think it should be doing. Therefore, causality does not apply to systems that are absolutely small. The act of observation destroys causality. Indeterminacy, that is a certain type of indefiniteness, comes into the calculation of observables for absolutely small systems. A system is absolutely small if the minimum disturbance that accompanies a measurement is not negligible. An absolutely small system can't be observed without changing it.

CAN'T CALCULATE THE
FUTURE—ONLY PROBABILITIES

Unlike in classical mechanics, once an observation is made for a quantum system, it is not possible to say exactly what another observation will yield. This lack of exactitude is not like the bird hitting the rock in Figure 2.6. In the bird-rock case, it is possible, if difficult, in principle to predict the result of the next observation. We would need to know all of the properties of the bird and the rock, as well as the exact details of how the bird hit the rock (e.g., the velocities and masses of the bird and the rock and the angle at which they hit). In the electron–photon case, it is impossible to predict exactly what the results of the next observation will be. What quantum theory can do is predict the probability of obtaining a particular result. In the Schrödinger's Cats example, when a box was opened, either a dead cat or a live cat was found. There was no way to predict which it would be. Opening the box (observing the cat) changed the cat from being in a type of mixed live-dead state into either a pure live state or a pure dead state. If many boxes were opened, the probability of finding a cat alive was 50%, but there was no way to predict what would happen when a particular box was opened (a single measurement). The cat problem is not physically realizable and, therefore, it is not a true quantum mechanical problem. A physically real problem that is akin to the cat problem is discussed in a later chapter. The cat problem was intended to introduce the idea that an observation can change a system and that only a probability could be ascertained from a series of experiments. For real systems that are absolutely small, quantum mechanics is the theory that permits the calculation and understanding of the distribution of probabilities that are obtained when measurements are made on many identically prepared systems. How quantum mechanical probability distributions arise and how to think about the nature of the disturbance that accompanies measurements on absolutely small systems are discussed in the following chapters.

3

Some Things About Waves

To address the nature of the inherent disturbance that accompanies a measurement and to understand what can and cannot be measured about an absolutely small quantum mechanical system, first it is necessary to spend some time discussing classical waves and the classical description of light. At the beginning of the twentieth century, a variety of experiments produced results that could not be explained with classical mechanics. The earliest of these involved light. Therefore, we will first discuss an experiment that seemed to show that classical ideas work perfectly. Then, in Chapter 4, we will present one of the experiments that demonstrated that the classical mechanics description could not be correct and, furthermore, that a classical reanalysis of the experiment seemed to work, but actually didn't. Finally, the correct analysis of the experiment involving light will be given using quantum ideas, which will bring us back to Schrödinger's Cat.

WHAT ARE WAVES?

There are many types of classical waves, water waves, sound waves, and light waves (electromagnetic waves). All waves have certain common properties, including amplitude, wavelength, speed, and direction of propagation (the direction in which a wave is traveling). Figure 3.1 shows a wave traveling in the x direction. The amplitude of the wave is the "distance" between its positive and negative peaks, the up-to-down distance. The wavelength is the distance along the direction of propagation between two positive or negative peaks. This is the distance over which the wave repeats itself. If you are riding on the wave and you move any integer number of wavelengths forward or backward along the wave, everything looks the same. The wave is traveling with some velocity, V.

WAVES HAVE VELOCITIES AND FREQUENCIES

The velocity depends on the type of wave, and the velocity of a wave needs a little discussion. Imagine you are standing beside the wave in Figure 3.1, but the wave is so long that you cannot see its beginning or end. Still, you can determine its velocity using a timing device. Start timing when a positive peak just reaches you and stop timing when the next positive peak reaches you. You now have enough information to determine the wave's velocity.

FIGURE 3.1. *A wave traveling in the x direction. The black line represents zero amplitude of the wave. The wave undergoes positive and negative oscillations about zero. The distance between the peaks is the wavelength. The wave is traveling along x with a velocity V.*

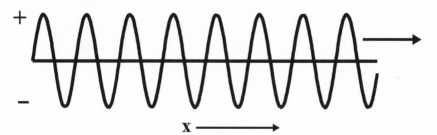

The wave has traveled d, a distance, of one wavelength, in time t. The distance equals the velocity multiplied by the time, d = Vt. (If you are in a car going at velocity, V = 60 miles per hour, and you travel for a time, t = 1 hour, then you have traveled a distance, d = 60 miles.) If we take the distance of one wavelength and divide it by the time it took to travel one wavelength, then we know the velocity, V = d/t. Watching the wave go by is like watching a very long train go by. You see boxcar after boxcar pass you. If you know the length of one boxcar and how long it takes that one boxcar to pass by, then you can determine the velocity of the train.

Another important property of waves that is related to their velocity and wavelength is the frequency. Scientists love using Greek letters to represent things because we tend to use up all of the Roman letters early on. There is no reason the velocity has to be V or distance d or time, t, but these are usually used, and many of the letters of the Roman alphabet have common usages. Therefore, we turn to the Greek alphabet. It is common to call a wave's wavelength λ (lambda) and a wave's frequency ν (nu). To see what the frequency is, again consider the train of box cars passing by. If you count how many boxcars go by in a certain amount of time, you have found the box car frequency. If 10 boxcars go by in a minute, the frequency is 10 per minute, which would usually be written as 10/minute. The frequency of a wave is determined by how many cycles (peaks) go by a point in a second. If 1000 cycles pass by a point in a second, the frequency is $\nu = 1000/s = 1000$ Hz. Lowercase s is used for the units of seconds. Per second has its own unit, Hz, for Hertz, which is in honor of Gustav Ludwig Hertz (1887–1975), who shared the Nobel Prize in Physics in 1925 with James Franck "for their discovery of the laws governing the impact of an electron upon an atom." The wavelength, velocity, and frequency of a wave are related through the equation, $\lambda\nu = V$.

OCEAN WAVES

Waves in the deep ocean travel with the crest above the average sea level and the troughs below sea level. A typical ocean wave has a wavelength $\lambda = 160$ m (520 ft) and travels with a velocity of 60 km/hr (60 kilometers per hour, or 38 miles per hour). The period, which is the time between wave crests, is 10 s, so the frequency $v = 0.1$ Hz. The amplitude is just the distance between a crest and a trough. Therefore, it is relatively straightforward to visualize the amplitude. (Waves break at the beach because the troughs drag on the ocean bottom in shallow water, which slows them down. The crests move faster than the troughs and fold over to produce the breaking waves we see at the beach. Waves traveling in the ocean do not break.)

SOUND WAVES

Sound waves are density waves in air. A standard tuning fork A above middle C is 440 Hz. When you strike the tuning fork, the tines vibrate at 440 Hz. The vibration produces sound waves. The tines moving back and forth "push" the air back and forth at 440 Hz, producing a wave with frequency, $v = 440$ Hz. At 70°F, the speed of sound is $V = 770$ miles per hour, which is 345 m/s. Because $\lambda v = V$, the wavelength of the 440 Hz sound wave is $\lambda = 0.78$ m (2.55 ft). The sound wave consists of air density going above the average density and then below the average density, more air and then less air. The density is the weight of air in a unit of volume, for example the number of grams in a cubic centimeter (g/cm^3). Increased density can be associated with increased pressure. So you could also think of the sound wave as a pressure wave in which the air pressure goes up and down at 440 Hz. When the sound wave enters your ear, the up-and-down oscillation of the pressure causes your eardrum to move in and out at the frequency of the sound wave, in this case, 440 Hz. The motion of the eardrum transfers the sound into the interior of the ear and tiny hairs are wiggled depending on the frequency of the sound. The motion of these hairs stimulates nerves, and the brain decodes the nerve impulses into what we perceive as sounds.

The amplitude of a sound wave is the difference between the maximum and minimum density (maximum and minimum pressure). In contrast to an ocean wave, you cannot see the amplitude of a sound wave, but you can certainly hear the differences in the amplitudes of sound waves. It is relatively simple to obtain electrical signals from sound waves, which is what a microphone does. Once an electrical signal is produced from a sound wave, its amplitude can be measured by measuring the size of the electrical signal. Like all classical waves, sound waves propagate in a direction and have an amplitude, a wavelength, and a velocity.

CLASSICAL LIGHT WAVES

The discussion of ocean waves and sound waves sets the stage for the classical description of light as light waves. In the classical description of light, explicated in great detail with Maxwell's Equations (James Clerk Maxwell, 1831–1879), light is described as an electromagnetic wave. The wave has an electric field and a magnetic field, both of which oscillate at the same frequency. You have experienced electric and magnetic fields. If you have seen a magnet pull a small object to it, then you have seen the effect of a magnetic field. The magnetic field from a magnet is static, not oscillatory as in light. You may have also seen the effects of electric fields. If you have combed your hair on a very dry day with a plastic comb, you may have noticed that your hair is attracted to it. After combing, very small bits of paper may jump to the comb as the comb is brought close to them. These effects are caused by a static electric field. An electromagnetic wave has both electric and magnetic fields that oscillate.

Unlike ocean waves, which travel in water, and sound waves,

which travel in air, light waves can travel in a vacuum. In a vacuum, the velocity of light is given the symbol c, and $c = 3 \times 10^8$ m/s. The speed of light is about a million times faster than the speed of sound. This is the reason why you see distant lightning long before you hear it. Sound takes about 5 seconds to travel a mile. Light takes about 0.000005 s or 5 μs (microseconds) to travel a mile. The velocity of light is slower when it is not traveling in a vacuum. In air it is almost the same as in a vacuum, but in glass it travels at about two-thirds of c.

What is an electromagnetic wave, which is the classical description of light? In a water wave, we have the height of the water above and below sea level oscillating. In a sound wave, the air density or pressure oscillates above and below the normal values. If you take a small volume, the amount of air (number of molecules that make up air, mostly oxygen and nitrogen) goes above and below the average amount of air in the volume. In an electromagnetic wave, two things actually oscillate, an electric field and a magnetic field. We usually talk about the electric field because it is easier to measure than the magnetic field. The oscillating electric field is an electric wave. When you listen to the radio, the radio antenna is a piece of wire that detects the radio waves. Radio waves are just low frequency electromagnetic waves. They are the same as light waves, but much lower in frequency. The electric field in an electromagnetic wave oscillates positive and negative from a maximum positive amplitude value to the same negative value. The metal in a radio antenna has many electrons that can be moved by an electric field. (Electrons will be discussed in detail further on, and electrical conduction will be discussed in Chapter 19.) The oscillating electric field of a radio wave causes the electrons in the antenna to oscillate back and forth. The electronics in the radio amplify the oscillations of the electrons in the antenna and convert these oscillations into an electrical signal that drives the speakers to make the sound waves that you hear. So we can think of light classically as an oscillating electric field and an oscillating magnetic field. Both oscillate at the same frequency and travel together at the same speed in the same direction. This is why they are called electromagnetic waves.

VISIBLE LIGHT

For light in a vacuum, $\lambda\nu = c$. The visible wavelengths, that is, the wavelengths we can see with our eyes, range from 700 nm (red) to 400 nm (blue). (A nm is a nanometer, which is 10^{-9} meters or 0.000000001 meters.) The visible wavelengths of light are very small; the velocity of light is very high. Therefore, the frequencies of visible light waves are very high. Red light has $\nu = 4.3 \times 10^{14}$ Hz, and blue light has $\nu = 7.5 \times 10^{14}$ Hz. 10^{14} is 100 trillion. Contrast light frequencies to a sound wave frequency (440 Hz) or an ocean wave frequency (0.1 Hz). Unlike an ocean wave or a sound wave, there is a complication in measuring the amplitude of a light wave. The frequency of light is so high that even the most modern electronics cannot see the oscillations. Rather than measuring the amplitude of the wave, defined as the amplitude of the oscillating electric field, the intensity of light is measured. The intensity, I, is proportional to the absolute valued squared of the electric field E, which is written as $I \propto |E|^2$. The absolute value, the two vertical lines | |, just means, for example, if there is a sign, positive or negative, we ignore it and just make everything positive. A photodetector, like the CCD in a digital camera (a CCD, or charge coupled device, makes an electrical signal when light strikes it), measures the amount of light, the intensity, rather than the amplitude of a light wave. Your eye does not directly measure the frequency of light waves in contrast to your ear, which measures the frequency of sound waves.

ADDING WAVES TOGETHER—INTERFERENCE

Waves of any kind, including light waves, can be added together to give new waves. Figure 3.2 shows on the left two identical waves (same wavelength, same amplitude, propagating in the same direction) that are in phase. (The waves are actually on top of each other, but they have been displaced so that we can see them individually.) "In phase" means that the positive peaks of one wave line up exactly with the positive peaks of the other wave, and therefore, the negative peaks also line up. The vertical dashed line in Figure 3.2 shows that the peaks line up. When waves are in phase, we say that the phase difference is 0° (zero degrees). One cycle of a wave spans a phase of 360°. Starting at any point on a wave, if you go along the wave for 360°, you are in an equivalent position, like going 360° around a circle. When two identical waves are added in phase, the resultant wave has twice the amplitude. This is called constructive interference, as shown on the right side of Figure 3.2.

Waves that are 180° out of phase can also be added together. As shown on the left side of Figure 3.3, waves that are 180° out of phase have the positive peaks of the top wave exactly lined up with the negative peaks of the bottom wave and vice versa. (Again, for interference to occur the waves need to actually be on top of each other, but they have been displaced so that we can see them clearly.) The dashed vertical line in Figure 3.3 shows that the positive peak of one wave is exactly lined up with the negative peak of the other

FIGURE 3.2. *Two identical waves that are in phase. The waves undergo positive and negative oscillations about zero (horizontal line). The positive peaks line up, and the negative peaks line up. They undergo constructive interference (are added together) to form a wave with twice the amplitude.*

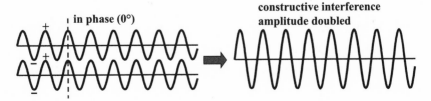

FIGURE 3.3. *Two identical waves that are 180° out of phase. The waves undergo positive and negative oscillations about zero (horizontal line). The positive peaks of the top wave line up exactly with the negative peaks of the bottom wave, and the negative peaks of the top wave line up exactly with the positive peaks of the bottom wave. The two waves undergo destructive interference when they are added together to produce zero amplitude.*

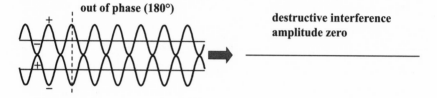

wave. When two identical waves that are 180° out of phase are added, the positive peaks and the negative peaks exactly cancel. For example, take the maximum positive value to be +1 and the maximum negative value to be −1. Adding +1 and −1 gives zero. In Figure 3.3 each point on the top wave that is positive lines up perfectly with a point on the bottom wave that is the same amount negative, and each point of the top wave that is negative lines up with an equivalent point on the bottom wave that is the same amount positive. Therefore, the waves exactly cancel to give zero amplitude as shown on the right side of the figure. This cancellation is called destructive interference.

INTERFERENCE PATTERNS AND THE OPTICAL INTERFEROMETER

Waves do not have to be right on top of each other and going in the same direction to interfere. They just have to overlap in some region of space, and interference can occur in that region. When Davies

Symphony Hall in San Francisco was opened in 1980, it had acoustic problems. While the problems were very complicated, it is easy to see how they developed. Imagine that you are sitting in the audience pretty far back from the orchestra. When a 440 Hz A is played, the acoustic wave comes directly at you but it also bounces off of the walls on either side of you. If there is a reflection from the wall to your right and a reflection from the wall to you left so that the reflected acoustic waves (sound waves) from each wall comes to your row of seats at, for example, a 30° angle, an interference pattern will be produced along your row of seats. There will be places where reflected waves constructively interfere and make the sound louder and places where the waves destructively interfere and make the sound softer. The spacing between a peak and a null of the interference pattern is 2.4 ft (see below for the spacing formula). So depending on your seat, the 440 Hz A will be louder or softer. Of course, there are many frequency acoustic waves coming at you from many directions. The combined interference effects distorted the sound that should have been coming straight at you from the orchestra. The problem in Davies Hall was fixed in 1992 by the installation of 88 carefully designed panels hanging from the ceiling along the two side walls. No two panels are identical. They are filled with sand and weigh as much as 8500 pounds. These panels prevented the reflections from the walls from going into the audience.

Light can also undergo interference phenomena. The classical view of optical interference patterns can reproduce experimental results, as we are about to see. However, as discussed in Chapters 4 and 5, ultimately the classical description fails when other experiments are considered. The correct description will introduce the quantum mechanical superposition principle and bring us back to Schrödinger's Cats.

Figure 3.4 shows a diagram of an interferometer used by Michelson (Albert Abraham Michelson, 1853–1931) in his studies of

FIGURE 3.4. *The incoming light wave hits a 50% reflecting mirror. Half of the light goes through the mirror and half reflects from it. The light in each leg of the interferometer reflects from the end mirrors. Part of each beam crosses in the overlap region at a small angle. To the right of the circled overlap region is a blowup of what is seen along the x direction when two beams cross. An interference pattern is formed in which the intensity varies along x from a maximum value to zero periodically.*

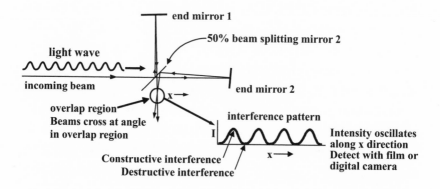

the nature of light waves. Michelson won the Nobel Prize in Physics in 1907 "for his optical precision instruments and the spectroscopic and metrological investigations carried out with their aid." Michelson and Morley, who was a coworker of Michelson, used an interferometer to attempt to determine the nature of the medium in which light waves propagated. Water waves propagate in water. Sound waves propagate in air. The Michelson-Morley experiment showed that light waves do not have an underlying medium, which had been called the aether. Light can propagate in a vacuum. There is no aether that pervades space. Light waves traveling to us from the stars are not traveling in a medium the way ocean waves and sound waves travel in water and air, respectively. This was an important step in recognizing that light waves are not waves in the same sense as sound waves. Here we only want to understand the classical description of what is observed with an interferometer.

In Figure 3.4, a beam of light, taken to be a light wave, enters the apparatus from the left. The light hits a partially reflecting "beam-splitting" mirror that reflects 50% of the light intensity and transmits 50% of the light intensity. In the wave description of light, there is no problem having part of the wave go one way and part the other way. The reflected light goes vertically up the page, reflects from end mirror 1, which is at a small angle so the reflected beam does not quite go right back along the same path. The reflected beam goes down the page and part of it goes right through the beam-splitting mirror. (Part of this beam reflects from the beam splitter, but we are not concerned with this portion.) This path is leg 1 of the interferometer. The 50% of the original beam that goes through the beam splitter hits end mirror 2, which is also at a small angle. This reflected beam travels back to the left, almost retracing its original path. It reflects from the beam splitter. (The portion that goes through the beam splitter is unimportant for our considerations.) The reflected portion heads down the page. This path is leg 2 of the interferometer. The result is that the two beams, one that traveled leg 1 and one that traveled leg 2, come together after traveling the same distance and cross at a small angle in the "overlap region" shown by the circle in Figure 3.4. This crossing of the light waves is like the crossing of the sound waves in Davies Symphony Hall that caused the interference problems.

In Figure 3.4 the light beams are drawn as lines, but in any real experiment the beams have a width. The x direction shown in the figure is perpendicular to the bisector of the angle (the line that splits the angle) made by the crossing beams. Since the angle is small, the x direction is basically perpendicular to the propagation direction of the beams, and in the figure it is the horizontal direction. A blowup of what is seen along the x direction in the overlap region is shown in the lower right portion of the figure. In the graph the vertical axis is the intensity of the light, I, and the horizontal axis is the position along x. Because the beams cross at a small angle, the phase relationship between them varies along the x direction, and there are alternating regions of constructive and destructive interference. The intensity of the light varies from a maximum value to zero back to the maximum, and again to zero, and so on. The crossed light waves form regions of constructive and destructive interference. At the intensity maxima, the light waves are in phase (0°—see Figure 3.2), and they add constructively to give increased amplitude. At the zeros of intensity, the light waves are 180° out of phase (see Figure 3.3), and they add destructively, to exactly cancel. This pattern can be observed by placing a piece of photographic film or a digital camera in the overlap region to measure the intensity at the different points along the x direction.

For a small angle, the fringe spacing, that is, the spacing, d, between a pair of intensity peaks or nulls is given by $d = \lambda/\theta$, where λ is the wavelength of light, and θ is the angle between the beams in radians (1 radian = 57.3 degrees). If 700 nm red light is used, and the angle between the beams is 1°, the fringe spacing is 40 μm or 1.6 thousandths of an inch. These fringes can be seen with film or a good digital camera. If the angle is 0.1°, the fringe spacing is 0.4 mm, which you can see by eye. If the angle is 0.01° (an exceedingly small angle), the fringe spacing is 4 mm (about a sixth of an inch), which you can easily see by eye. To have 4 mm fringes, the beams that cross must be much larger in diameter than 4 mm.

As discussed, in the classical description, light is an electromagnetic wave, and the intensity is proportional to the square of the electric field amplitude (size of the wave in Figure 3.1). In the following, we are not going to worry about units. By including a lot of constants, the units in the following all work out, but they are unimportant for our purposes here. Take the electric field in one of the beams in one leg of the interferometer to have an amplitude of 10. Then the intensity is 100 ($10^2 = 100 = 10 \times 10$). The other beam also has I = 100. These are the intensities when we are not observing in the beam overlap region. When the beams are separated, the sum of their intensities is 200. What happens in the overlap region? Waves constructively interfere in some places and destructively interfere in others (see Figure 3.4, lower right). Therefore, to determine the intensities in the overlap region, it is necessary to add the electric field amplitudes and then square the result to find the intensities. At an intensity maximum in the overlap region, the waves are perfectly in phase and add constructively. The electric field from beam 1 adds to the electric field from beam 2, that is, E = 10 + 10 = 20. Then the intensity in a peak in the interference pattern is $I = E^2 = 20^2 = 400$. The intensity is 400, twice as great as the intensity of just the sum of the intensities of the two beams by themselves when they are not constructively interfering. In a null of the interference pattern, the waves destructively interfere perfectly. An electric field of + 10 adds to an electric field of − 10, to give zero. The electric field equals zero, and I = 0. Therefore, the interference pattern is caused by alternating regions of constructive and destructive interference of electromagnetic waves. In some places the waves add, and we see a peak. In some places the waves subtract to give zero. Interference is a well-known property of waves, and the interference pattern produced by the interferometer seemed to be a perfect example of a wave phenomenon.

The interferometer and the interference pattern shown in Figure 3.4 can be described in complete detail using classical electromagnetic theory. The details of the interference pattern can be calculated with Maxwell's equations. This and many other experiments, including the transmission of radio waves, can be described with classical theory. Therefore, classical theory, which treats light as a wave, appeared to be correct up to the beginning of the twentieth century. However, Chapter 4 shows how Einstein's explanation of one phenomenon, the photoelectric effect, caused the beautiful and seemingly infallible edifice of classical electromagnetic theory to require fundamental rethinking.

4

The Photoelectric Effect and Einstein's Explanation

AT THE END OF THE NINETEENTH CENTURY, classical electromagnetic theory was one of the great triumphs of classical mechanics. It was capable of explaining a wide variety of experimental observations. But early in the twentieth century, new experiments were causing problems for the classical wave picture of light. One experiment in particular, along with its explanation, showed a fundamental problem with the seemingly indestructible wave theory of light.

THE PHOTOELECTRIC EFFECT

The experiment is the observation of the photoelectric effect. In the photoelectric effect, light shines on a metal surface and, under the right conditions, electrons fly out of the metal. For our purposes here, electrons are electrically charged particles. The electron charge is negative. (Later we will see that electrons are not strictly particles for the same reason that light is not a wave.) Because electrons are charged particles, they are easy to detect. They can produce electrical signals in detection equipment. Figure 4.1 shows a schematic of the photoelectric effect with the incoming light viewed as a wave.

It is possible to measure the number of electrons that come out of the metal and their speed. For a particular metal and a given color of light, say blue, it is found that the electrons come out with a well-defined speed, and that the number of electrons that come out depends on the intensity of the light. If the intensity of light is increased, more electrons come out, but each electron has the same speed, independent of the intensity of the light. If the color of light is changed to red, the electron speed is slower, and if the color is made redder and redder, the electrons' speed is slower and slower. For red enough light, electrons cease to come out of the metal.

THE WAVE PICTURE DOESN'T WORK

The problem for classical theory with these observations is that they are totally inconsistent with a wave picture of light. First, consider the intensity dependence. In the wave picture, a higher light intensity means that the amplitude of the wave is larger. Anyone who

FIGURE 4.1. *The photoelectric effect. Light impinges on a metal, and electrons (negatively charged particles) are ejected. In the classical picture, light is a wave, and the interaction of the wave with the electrons in the metal causes them to fly out.*

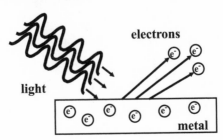

has been in ocean waves knows that a small wave hits you gently and a big wave hits you hard. As illustrated in Figure 4.2, low-intensity light is an electromagnetic wave with small amplitude. Such a wave should "hit" the electrons rather gently. The electrons should emerge from the metal with a relatively slow speed. In contrast, high intensity light has associated with it a large amplitude wave. This large amplitude wave should "hit" the electrons hard, and electrons should fly away from the metal with a high speed.

To put this more clearly, the light wave has associated with it an oscillating electric field. The electric field swings from positive to negative to positive to negative at the frequency of the light. An electron in the metal will be pulled in one direction when the field is positive and pushed in the other direction when the electronic field is negative. Thus, the oscillating electric field throws the elec-

FIGURE 4.2. *A wave picture of the intensity dependence of the photoelectric effect. Low-intensity light has a small wave amplitude. Therefore, the wave should "hit" the electrons gently, and they will come out of the metal with a low speed. High-intensity light has a large wave amplitude. The large wave should hit the electrons hard, and the electrons will come out of the metal with a high speed.*

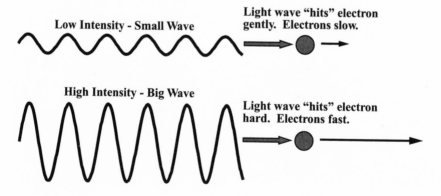

tron back and forth. According to classical theory, if the wave has large enough amplitude, it will throw the electron right out of the metal. If the amplitude of the wave is bigger (higher intensity), it will throw the electron out harder, and the electron should come out of the metal having a greater speed. However, this is not what is observed. When the intensity of light is increased, electrons come out of the metal with the same speed, but more electrons come out.

Furthermore, when the light color is shifted to the red (longer wavelength), the electrons come out of the metal with a lower speed no matter how high the intensity is. Even in the wave picture, longer wavelength light is less energetic, but it should be possible to turn up the intensity, making a bigger amplitude wave, and therefore increase the speed of the electrons that fly out of the metal. But, as with a bluer wavelength, turning up the intensity causes more electrons to emerge from the metal, but for a given color, they all come out moving with the same speed.

An additional problem is that when the color is shifted far enough to the red, the electrons stop coming out. The electrons have some binding energy to the metal, that is, the negatively charged electrons are attracted to the positively charged metal atom nuclei. (Atoms will be discussed in detail beginning in Chapter 9 and metals in Chapter 19.) This binding energy is what keeps the electrons from flying out of the metal in the absence of light. In the wave picture, it should always be possible to turn up the intensity high enough, and therefore make the amplitude of the oscillating electric field large enough, to overcome the binding energy. If you are standing in the ocean, a small wave may not knock you off of your feet, but if the waves get bigger and bigger, eventually they will be big enough to break the binding of your feet to the ocean floor and send you flying. But with light, for a red enough color, no matter how big the wave is, the binding of the electrons to the metal is not overcome.

EINSTEIN GIVES THE EXPLANATION

The upshot of these experimental observations is that the wave picture of light that describes the interference pattern of Figure 3.4 so well does not properly describe the photoelectric effect. The explanation for the photoelectric effect was given by Einstein in 1905 (Albert Einstein, 1879–1955). Einstein won the Nobel Prize in Physics in 1921 "for his services to Theoretical Physics, and especially for his discovery of the law of the photoelectric effect." It may seem surprising that Einstein, known for his Theory of Relativity, won the Nobel Prize for explaining the photoelectric effect, which was an important step in the transition from classical to quantum theory. Einstein's prize demonstrates the importance of the explanation of the photoelectric effect in modern physics.

Einstein said that light is not composed of waves, but rather of photons or quanta of light. In the photoelectric effect, a photon acts like a particle rather than a wave. So Einstein said that a beam of light is composed of many photons, each of which is a discrete particle. (As discussed in detail later, these are not particles in the classical sense of a particle.) As shown in Figure 4.3, one photon "hits" one electron and ejects it from the metal. The process is in

FIGURE 4.3. *Einstein described light as composed of discreet quanta of light "particles" called photons. In the photoelectric effect, one photon hits one electron and knocks it out of the metal.*

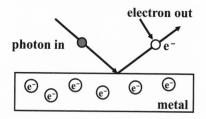

some sense like the cue ball in a game of pool hitting a stationary ball and sending it across the table. The cue ball hitting the stationary ball transfers energy to it in the form of kinetic energy, that is, energy of motion. The collision causes the cue ball to give up energy and the target ball to gain energy. A light beam is composed of many photons, but one photon ejects one electron from the metal.

When the intensity of light is increased, the light beam is composed of more photons. As illustrated in Figure 4.4, more photons impinging on the metal can hit and eject more electrons from the metal. Because one photon hits one electron, increasing the intensity of the light beam does not change the speed of the electron that is ejected. In pool, the speed of a target ball is determined by how fast the cue ball was moving. Imagine two cue balls were simultaneously shot at the same speed at two different target balls. After being hit, the two target balls would move with the same speed. When more photons of a particular color hit the metal, more electrons come out, but all with the same speed. In contrast to the wave picture, increased intensity does not produce a harder hit on an electron; increased intensity only produces more photons hitting more

FIGURE 4.4. *An increase in the intensity of a light beam corresponds to the beam being composed of more photons. More photons can hit and eject more electrons, so an increase in intensity results in more electrons flying out of the metal.*

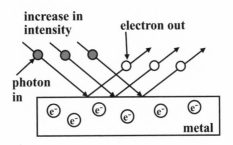

electrons. Each photon hits an electron with the same impact whether there are many or few. Therefore, electrons come out with the same speed independent of the intensity.

RED LIGHT EJECTS SLOWER
ELECTRONS THAN BLUE LIGHT

To explain why changing the color of the light to red (longer wavelength, lower energy) caused electrons to be ejected with a lower speed, Einstein used a formula first presented by Planck (Max Karl Ernst Ludwig Planck, 1858–1947). Planck first introduced the idea that energy comes in discreet units, called quanta, while he was explaining another phenomenon involving light, called black body radiation. When a piece of metal or other material is heated to a high temperature it will glow; it is emitting light. If it is quite hot, it will glow red. An example is the heating element of an electric stove or space heater when turned to high. As its temperature is increased, the color shifts toward blue. This is not only true of a piece of metal but also of stars. Red stars are relatively cool. A yellow star, such as our own sun, is hotter. A blue star is very hot. In 1900, classical physics could not explain the amount of light that came out at each color from a hot object. Planck found the explanation that still stands today by introducing a new concept, that the electrons in a piece of metal could only "oscillate" at certain discreet frequencies. The energy steps between these frequencies are called quanta. Planck won the Nobel Prize in Physics in 1918 "in recognition of the services he rendered to the advancement of Physics by his discovery of energy quanta." Planck's discovery of energy quanta led to the name Quantum Mechanics.

In his work, Planck introduced the formula that related the frequency of the electrons to their energy, $E = h\nu$, where ν is the frequency as discussed in Chapter 3, and h is called Planck's constant. In the equation, $h = 6.6 \times 10^{-34}$ J-s, J is the unit of energy

Joule, and s is seconds. In the formula, the units of v are Hz or 1/s; so h times v gives the units of energy, J. In his description of black body radiation, Planck postulated that the energy E could only change in discreet steps. It could be hv, of 2hv, or 3hv, and so on, but the energy could not have values between these integer step changes. The recognition that energy changes in discreet quanta at the atomic level marked the beginning of quantum mechanics.

Einstein proposed that Planck's formula also applied to photons, so that the energy of a photon was determined by its frequency v as E = hv. Using this formula, Einstein explained the reason that red light generates slower electrons than blue light. Red light is lower frequency than blue light. Therefore a red photon is lower energy than a blue photon. In the pool ball analogy, a blue photon hits the electron harder than a red photon, and therefore, the blue photon produces an electron that has a higher speed than a red photon. With this picture, it is clear why using redder and redder light produces slower and slower electrons emerging from the metal.

VERY RED LIGHT DOES NOT EJECT ELECTRONS

The one observation left to explain is why do the electrons stop coming out of the metal when the light is tuned far enough to the red? Einstein resolved this as well. When an electron is ejected from a metal by a photon, it has a certain kinetic energy. Kinetic energy means the energy associated with its motion. The higher the energy, the faster the electron moves. The kinetic energy is E_k where the subscript k stands for kinetic. The formula for kinetic energy is given by $E_k = \frac{1}{2}mV^2$, where m is the mass and V is the velocity. Then the velocity of an electron that emerges from a metal is related to its energy, which in turn is related to the energy of the photon that knocked it out of the metal. A higher energy photon will give the electron more kinetic energy, and the electron will move faster (have a larger V). As mentioned, electrons are held in a metal by a binding energy, call it E_b, where the subscript b stands for binding. Therefore, some of the energy that is carried by the photon has to go into overcoming the binding energy. The kinetic energy of the electron that comes out of the metal is just the photon energy, $E = h\nu$, minus the binding energy, E_b. Thus, the electron's kinetic energy is $E_k = h\nu - E_b$. For an electron to be ejected from the metal, the photon energy $h\nu$ must be larger than the binding energy E_b. As the light is tuned further and further to the red (longer wavelength, λ), ν becomes smaller and smaller because $\nu = c/\lambda$, where c is the speed of light. At some red enough color, $h\nu$ is less than E_b, and electrons are no longer ejected from the metal. Turning up the intensity causes more photons to impinge on the metal, but none of these photons has enough energy to eject an electron.

The fact that electrons stop coming out of the metal when the photon is tuned far enough to red (has low enough energy) can be understood by thinking of the child's game, Red Rover. In Red Rover, a line of kids stands across a field holding hands. A kid on the other team runs at the line. If he runs very fast (high energy), he breaks through the line and keeps going, although he is slowed down. If he runs somewhat slower, he will still break through the line. However, if he runs slow enough, he will not break through the line because his energy is insufficient to overcome the binding energy of the hands holding the line together.

HOW FAST IS AN EJECTED ELECTRON

It is interesting to get a feel for how fast an electron moves when it is ejected from a piece of metal. Different metals have different binding energies called work functions. A binding energy for a metal can be determined by tuning the color redder and redder and seeing the wavelength of light at which photons cannot eject electrons. For a metal with a small binding energy, a typical cutoff wavelength for electron ejection is 800 nm. For λ = 800 nm, ν = 3.75 $\times$ 10¹⁴ Hz, and E_b = $h\nu$ = 2.48 $\times$ 10⁻¹⁹ J. If we shine green light on the metal with a wavelength of 525 nm, the energy of the photon is 3.77 $\times$ 10⁻¹⁹ J. The kinetic energy of the electron that will be ejected from the metal is E_k = $h\nu - E_b$ = 1.30 $\times$ 10⁻¹⁹ J. We can find out how fast the electron is moving using $E_k = \frac{1}{2}m_e V^2$ = 1.30 $\times$ 10⁻¹⁹ J, where m_e is the electron mass, m_e = 9.11 $\times$ 10⁻³¹ kg (kg is kilograms, that is, 1000 grams). Multiplying the equation for E_k by 2 and dividing by m_e gives V^2 = 2(1.30 $\times$ 10⁻¹⁹ J)/m_e = (2.60 $\times$ 10⁻¹⁹ J)/(9.11 $\times$ 10⁻³¹ kg) = 2.85 $\times$ 10¹¹ m²/s². This value is the square of the velocity. Taking the square root, V = 5.34 $\times$ 10⁵ m/s, which is about one million miles per hour. In this example of the photoelectric effect, the ejected electrons are really moving.

Classical electromagnetic theory describing light as waves seems to work perfectly in describing a vast array of phenomena including interference, but it can't come close to explaining the photoelectric effect. Einstein explains the photoelectric effect, but now light can't be waves, so what happens to the classical description of interference? Reconciling the photoelectric effect and interference brings us to the cusp of quantum theory and back to Schrödinger's Cats.

5

Light:
Waves or Particles?

THE EXPLANATION OF THE photoelectric effect discussed in Chapter 4 required a new theoretical description of the interferometer experiment discussed in connection with Figure 3.4. Understanding the interferometer experiment in a manner that does not contradict the description of the photoelectric effect requires the big leap into thinking quantum mechanically rather than thinking classically. In discussing absolute size in Chapter 2, the idea was introduced that for a system that is small in an absolute sense, a measurement will make an unavoidable nonnegligible disturbance. However, we did not discuss the nature or consequences of such a disturbance. Now, we need to come to grips with the true character of matter and what happens when we make measurements.

The problem we have is that light waves were used to explain the interference phenomenon in Figure 3.4, but "particles of light," quanta called photons, were used to explain the photoelectric effect in connection with Figures 4.3 and 4.4. The classical mathematical description of light waves employed Maxwell's equations to quantitatively describe interference. The mathematical entity that represented a light wave in the theory is called a wavefunction. A function gives a mathematical description of something, in this case a light wave. It describes the amplitude, frequency, and the spatial location of a light wave. The incoming light wave is described by a single wavefunction. In the classical description, after the light wave hits the 50% beam splitter, half of the wave goes into each leg of the interferometer (see Figure 3.4). There are now two waves and two wavefunctions, one for each wave. These wavefunctions describe waves that are each half of the intensity of the original incoming wave and are in different locations, the two legs of the interferometer. When these two wavefunctions are combined mathematically to describe the nature of the overlap region inside the circle in Figure 3.4, the interference pattern can be calculated. All of this worked so well that it was thought that the same math must apply to photons.

CLASSICAL DESCRIPTION OF INTERFERENCE DOESN'T WORK FOR PHOTONS

Figure 5.1 shows the interferometer again with everything exactly the same as in Figure 3.4 except that the incoming light beam is now composed of photons. Initially, it was thought that when the beam of photons hits the 50% beam splitter, half of the photons go into leg 1 of the apparatus and head toward end mirror 1, while the other half of the photons go into leg 2 of the apparatus and head for end mirror 2. The photons would then reflect from the end mirrors and after again hitting the beam splitter, half of the photons from each leg would cross in the overlap region. The inference pattern was thought to develop when photons from one leg of the apparatus interfered with photons from the other leg of the apparatus. This thinking proved to be incorrect.

To describe the interference effect, the mathematical formulation in terms of Maxwell's wavefunctions was not changed at all.

FIGURE 5.1. *The beam of light is composed of photons that hit a 50% reflecting mirror. In the initial incorrect description of the interference effect in terms of photons, it was thought that half of the photons go into each leg of the interferometer. The photons from each leg cross in the overlap region, and it was believed that the photons from one leg interfere with photons from the other leg to produce the interference pattern. The idea that photons from one leg interfere with photons from the other leg is not correct.*

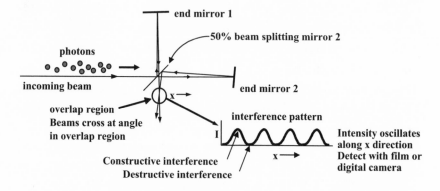

However, the physical meaning of the wavefunction was redefined. Instead of the amplitude of an electromagnetic wave in a certain region of space, such as leg 1 or leg 2 of the interferometer, the wavefunction was redefined as describing *the number of photons in a region of space.* Previously, the wavefunction was taken to give the amplitude of the wave in a region of space, and this amplitude could be used to calculate the intensity. After the redefinition, the wavefunction was taken to tell how many photons were in a region of space, say leg 1 of the interferometer, and the intensity could still be calculated. This redefinition seemed perfectly reasonable, but it was wrong! The entire description of half of the photons going into each leg of the interferometer is fundamentally wrong. The correct description required the leap to thinking quantum mechanically.

Many things are wrong with the picture that half of the photons go into each leg of the interferometer and then come together and interfere with each other. The simplest experiment that shows the problem with this description is the intensity dependence of the interference pattern (the lower right portion of Figure 5.1). The shape of the interference pattern observed in the overlap region of the interferometer is independent of the intensity used to create it. For a given method of detection, a piece of photographic film or a digital camera, note that if the intensity is turned up it will take less time to acquire a high-quality pattern, but the shape of the pattern is unchanged. That is, the separation and shapes of the peaks and nulls in the pattern are unchanged. As discussed in Chapter 3, the periodicity of the pattern depends on the crossing angle of the beams and the wavelength of light. It does not depend on the intensity. If the intensity is reduced, it will take more time to collect the pattern, but the pattern will not change shape. A standard red laser pointer puts out 1 mW (milliwatt), which is one thousandth of a watt, or 0.001 J/s (joules per second). The red color is approximately 650 nm, that is, the wavelength $\lambda = 650$ nm. Using $\lambda v = c$ and $E = hv$ where h is Planck's constant, v is the frequency of the light, and c is the speed of light, one photon of 650 nm light is about 3 $\times$ 10^{-19} J. So the 1 mW laser pointer is emitting about 3 $\times$ 10^{15} photons per second, that is 3 thousand trillion photons per second. If this is the input beam for the interferometer, the interference pattern will be very easy to record and, in fact, if the fringe spacing is big enough (see Chapter 3, the discussion of fringe spacing following Figure 3.4), you will be able to see the interference pattern with your eyes.

Imagine turning the intensity down and down. Soon, you will not be able to see the interference pattern because your eyes are not very sensitive light detectors, but you can still record it with photographic film or a digital camera. Once recorded, the pattern is the same. Turn the intensity down a factor of 3000 to a trillion photons a second, and the pattern is unchanged. In the description in which half of the photons go into each leg of the apparatus, half a trillion photons per second go into each leg of the interferometer. Turn down the intensity to a billion photons per second, and the pattern is the same. Further reduce the intensity to a million photons per second, and there is still no change. Here is where the fallacy in the description becomes obvious. Turn down the intensity until there is only one photon per second entering the apparatus. Again, the pattern is unchanged. At one photon per second, it will take a long time to record enough signal to see the interference pattern, but if you wait long enough the pattern is the same.

When only one photon per second is entering the interferometer, there is only one photon at a time in the apparatus. A photon will take on the order of one hundred millionth of a second (10^{-8} s) to traverse the interferometer. With one photon per second, there is virtually no chance that there is more than one photon at a time in the instrument, yet the interference pattern, once recorded, is the same. But the modified classical description of the interference effect in terms of photons said that half of the photons go into leg 1 and half of the photons go into leg 2. The photons in leg 1 interfere with the photons in leg 2 to produce the interference pattern. If there is only one photon in the apparatus at a time, there is no other photon for it to interfere with. The "half of the photons go into each leg of the apparatus model" predicts that the interference pattern should vanish at low enough intensity. The interference pattern does not disappear at low intensity. The model is wrong!

A NEW DESCRIPTION OF PHOTONS IN THE INTERFEROMETER

Here is where the complete change in thinking is required that will bring us back to Schrödinger's Cats. How is it possible to have an interference pattern when only one photon enters the interferometer at a time? Our understanding of this problem and the nature of quantum mechanics in general is based on the conceptual interpretation of the mathematical formalism that is strongly associated with the work of Max Born (1882–1970). Born won the Nobel Prize in Physics in 1954 "for his fundamental research in quantum mechanics, especially for his statistical interpretation of the wavefunction." This interpretation is frequently referred to as the Copenhagen interpretation.

The correct description of the interferometer experiment is that *each photon goes into both legs of the interferometer*. This is the big leap. A single photon encounters the 50% beam splitter. That means there is a 50% chance that the photon will be reflected and go into leg 1 of the interferometer (see Figure 5.1) and a 50% chance it will go into leg 2. Classically one would say the photon must go one way or the other, that is, it either goes into leg 1 or it goes into leg 2. This is not correct. When the photon encounters the beam splitter, its state is changed. If a photon is in fact moving in leg 1, call this state of motion "translation state 1," abbreviated T1. If a photon is moving in leg 2, call this state of motion "translation state 2," abbreviated T2. After a photon interacts with the beam splitter, it is not in T1 or T2. The state of the system after the beam splitter is referred to as a superposition state. It is an equal mixture of T1 and T2. In some sense, the photon is simultaneously in both T1 and T2. This sounds really strange. The single photon is in two regions of space simultaneously. It is in a superposition translation state $T = T1 + T2$. It is in a state that is an equal mixture of T1 and T2.

The photon is in the superposition translation state $T = T1 + T2$ because this is what is known about it. It has a 50% chance of being in leg 1 (T1) and a 50% chance of being in leg 2 (T2). The Born interpretation of the wavefunction states that the wave is not a real wave in the sense of the amplitude of an oscillating electromagnetic field. Rather, the wavefunction describes a "probability amplitude wave." The *incorrect interpretation* of the wavefunction in terms of photons is that it tells how many photons are in each leg of the apparatus, that is, how many photons are in a region of space. The *correct interpretation* is that the wavefunction is related to *the probability of finding a photon in a region of space*. The difference between the incorrect and correct interpretation may not seem like a major difference, but, as shown in detail below, it is a fundamental change in our view of nature. In the classical description of light, the intensity is proportional to the absolute value squared of the amplitude of the electric field, which in turn was given by the amplitude of the wavefunction. In the Born interpretation, the absolute value squared of the wavefunction for a certain region of space gives the probability of finding a particle, in this case a photon, in that region of space.

A PHOTON INTERFERES WITH ITSELF

When the photon meets the beam splitter, two probability amplitude waves are created, one in leg 1 and one in leg 2. The total probability amplitude wave T is the superposition of the probability amplitude waves T1 and T2. After encountering the beam splitter, each and every photon is in a state T1 + T2. Because there are two probability amplitude waves after the beam splitter, they cross in the overlap region. A single photon inside the interferometer has two waves, T1 and T2, associated with it. The interference of these two waves determines the probability of finding a photon near the peak, which is high, and finding a photon near the null, which is low. A photon interferes with itself because it is composed of two waves in the interferometer, and two waves can interfere with each other. Because every single photon is placed in the T1 + T2 superposition state after meeting the beam splitter, there is no problem with turning down the intensity. A single photon entering the apparatus produces two waves, probability amplitude waves, in the interferometer. Therefore, there is always a pair of waves to produce an interference pattern.

A PHOTON CAN BE IN TWO PLACES AT ONCE

The first natural response of a classical thinker to the Born interpretation is "this is nuts." Are we really to believe that a single photon can be in two places at once? After meeting the beam splitter, the state that is produced is T1 + T2. The state T1 + T2 means that in some sense the photon is simultaneously in both legs of the apparatus. If this is true, why don't we just make some measurements to see where the photon is? It doesn't do much good to make the measurement with trillions of photons going into the apparatus. If we put an instrument in leg 1 to see how much light there is, we will find half of the light. However, that doesn't tell us what we want to know. Maybe half of the photons go in each leg and we see half, or maybe there is a 50% chance that each photon goes into each leg. We will still see half. The correct experiment is to use such low-intensity light so that only one photon is in the apparatus at a time.

Consider the experiment in which we shoot one photon at a time into the interferometer. We use a photodetector that is so sensitive that it can detect a single photon. This is readily doable with the scientific equivalent of a superdigital camera. We place the detector in leg 1 of the interferometer. A photon enters the apparatus, and we detect it. We see an entire photon. We don't see half a photon. Another photon goes in the apparatus, and we don't see a photon. Five more photons enter the apparatus. We detect two of them and do not detect the other three. After doing this for a long time, it is found that 50% of the photons are observed by the detector placed in leg 1 of the apparatus. We also find that no interference pattern was produced. In fact, what is observed is a single bright spot (no oscillatory pattern) in the region in which the interference pattern occurred previously.

Observation Causes a Nonnegligible Disturbance Causing a Change of State

What is going on? After the photon meets the beam splitter, it is in a superposition state, T1 + T2. However, photons are particles that are small in the absolute sense. The act of making an observation causes a nonnegligible disturbance. When we place the photodetector in leg 1 of the apparatus, we are making an observation of the location of the photon. The act of making the observation causes the system to jump from being in the superposition state T1 + T2 into either a pure state T1 or a pure state T2. The superposition wavefunction has been "collapsed" into one of the pure states that make up the superposition. If the system makes the jump into the state T1, the photon is detected. Of course, once it slams into the photodetector, it does not continue to propagate through the interferometer. If the photon jumps into the state T2, it is not detected with the photodetector that is located in leg 1, and it continues to propagate, eventually reaching the region in which the instrumentation is set up to observe the interference pattern. However, because the photon is in the pure state T2, there is only a single probability amplitude wave. When it reaches the "overlap" region (see the bottom of Figure 5.1), there is no other probability amplitude wave to interfere with. Therefore, no interference pattern is created. A single spot is formed as each photon that traverses the apparatus in the pure T2 state hits a single spot on the detector like a bullet aimed at that spot. The spot has the same size (diameter) as the initial light beam that entered the apparatus, but no spatially oscillatory interference pattern.

BACK TO SCHRÖDINGER'S CATS

Making the observation of the location of the photon with the photodetector in leg 1 of the interferometer causes the photon to jump from the mixed superposition state T1 + T2 into a pure state, either T1 or T2. However, on a single measurement it is not possible to know which state will be produced by the observation. The chance is 50/50 that it is in T1 and 50/50 that it is in T2. After many measurements, we know that the probability of making the jump to T1 is 50%, but it is impossible to say in advance what will happen for any single observation. This is a true physical manifestation of the issues raised with Schrödinger's Cats in Chapter 1, where we had 1000 boxes with a cat in each one. Each cat was in a superposition state that was 50% alive and 50% dead. In this very nonphysical heuristic scenario, when a box is opened an observation is made as to the health of the cat. Sometimes the cat is perfectly healthy and sometimes the cat is dead. After opening all of the boxes, it was determined that the probability of finding a live cat was 50%, but there is no way of predicting ahead of time when opening a particular box, that is, on making a single observation, whether a live or dead cat will be found. Before opening the box, the cat is in a superposition state that is a 50/50 mixture of live and dead. The act of making the observation makes a nonnegligible disturbance and causes the superposition state to jump into either a pure live state or a pure dead state. As discussed in Chapter 1, a live/dead cat superposition state does not and cannot exist, but the interferometer is a real example of the ideas illustrated by Schrödinger's Cats.

A photon can easily be placed into a superposition state consisting of a 50/50 mixture of two translations states using the 50% beam splitter. When it is in the superposition state, it is not possible to say whether the photon is in leg 1 or leg 2 of the apparatus. It is only possible to say that if we make a measurement to see where the photon is, the measurement will make a nonnegligible disturbance. The disturbance will cause the state of the system to change from having equal probability of being in the two legs of the interferometer to being in either one or the other. The interference pattern is produced when the photon's probability amplitude waves interfere with each other. The two pieces of the superposition state, T1 and T2, which comprise the total probability amplitude wave for a photon in the interferometer, interfere with each other. If an observation is made to see where the photon is, it will be found to be either in leg 1 or leg 2 of the apparatus. However, the act of observation changes the system so that it is no longer in a superposition state. There are no longer two parts of the probability amplitude wave to interfere with each other, and the interference pattern vanishes. Thus, a photon in an interferometer is a real manifestation of the ideas relating to Schrödinger's Cats.

BACK TO THE PHOTOELECTRIC EFFECT

In Chapter 4, the photoelectric effect was described in terms of photons, which are particles that behave in some sense like bullets of light. One photon strikes one electron and knocks it out of a piece of metal (see Figure 4.3). The description of the photoelectric effect showed that the classical description of light as an electromagnetic wave was incorrect. A new concept had to be introduced to explain both the photoelectric effect and the fact that photons could produce an interference pattern. The Born interpretation of the wavefunction as a probability amplitude wave gave the photon the necessary wavelike characteristics, so that photons could produce an interference pattern. However, in discussing the probability amplitude waves in connection with the interferometer, we only considered the location of the photon in terms of two rather large regions of space; a photon was in a superposition state, T1 + T2, with equal probabilities of being in leg 1 and leg 2 of the interferometer. The photoelectric effect implied that a photon is quite small. Chapter 6 will show how a superposition of probability amplitude waves can produce a photon that is very small in size. The ideas will lead to one of the central and most nonclassical aspects of quantum mechanics, the Heisenberg Uncertainty Principle.

6

How Big Is a Photon
and the Heisenberg
Uncertainty Principle

IN CHAPTER 5, WE LEARNED that a photon in an interferometer interferes with itself. In some sense, a photon can be in more than one
place at a time. The photon location is described as a probability
amplitude wave. This is not like a water wave, a sound wave, or even
a classical electromagnetic wave. The wave associated with a photon
(or other particles like electrons) describes the probability of finding
the particle in some region of space. In the interferometer problem
(Figures 3.4 and 5.1), a single photon was in leg 1 and leg 2 simultaneously, with equal probability of finding the photon in either of
these regions of space. To understand and describe the location of
a photon in more detail, it is necessary to discuss more aspects of
waves. We need to know about the nature of the probability amplitude waves, particularly how they combine and what happens when
a measurement is made.

The simplest problem to discuss is a free particle, which was
introduced in Chapter 2. A free particle could be a photon, an electron, or a baseball. It is a free particle if no forces are acting on it.
That is, there is no gravity, no electric or magnetic fields, no photons

hitting an electron, no baseball bats hitting the baseball, no air resistance, and so on. With no forces acting on a particle, it has a perfectly
defined and unchanging momentum. Thus, if it is moving in a particular direction, it will just keep going in that direction. We can call
that direction anything we want, so let's call it the x direction. Think
of a graph with the horizontal axis x. We will just pick the direction
of the x axis to be along the direction the particle is moving. In connection with Figure 2.5, we talked about a classical particle moving
along x with a classical momentum p. Here we want to discuss the
nature of a quantum particle with momentum p.

PARTICLES HAVE WAVELENGTHS

For a photon, the momentum is given as $p = h/\lambda$, where h is Planck's constant and λ is the wavelength of the light. Therefore, the
momentum is related to the wavelength (the color) of the light.
Prince Louis-Victor Pierre Raymond de Broglie (1892–1987) won
the Nobel Prize in Physics in 1929 "for his discovery of the wave
nature of electrons." De Broglie showed theoretically that particles,
such as electrons or baseballs, also have a wave description. As discussed below, the wave description of electrons—or any type of particle—is in terms of the same types of waves as a photon, probability
amplitude waves, as introduced in Chapter 5.

The wavelength associated with a particle is $\lambda = h/p$. This is a
simple rearrangement of the formula for the photon momentum
given above. If both sides of the photon momentum formula are
multiplied by λ and divided by p, then the expression for the wavelength associated with a particle is obtained. De Broglie's important
result is that the relationship between the momentum and the
wavelength is the same for photons (light) as it is for material particles, such as electrons and baseballs. Therefore, the properties of
photons are described in fundamentally the same way as the properties of electrons, as well as baseballs. The wavelength associated

with a particle is called the de Broglie wavelength. (We will see with physical examples in the next chapter why baseballs don't seem to have wavelike properties, but photons and electrons do.)

WHAT A FREE PARTICLE WAVEFUNCTION LOOKS LIKE

For a free particle with some particular value of its momentum, p, what does the wavefunction look like? Recall that the wavefunction is related to the probability of finding the particle someplace in space. Figure 6.1 shows a graph of the wavefunction for a free particle with the momentum, p. As discussed above, the wavelength of the wavefunction associated with the particle is $\lambda = h/p$. As can be seen in the figure, the wavefunction for a free particle is represented by two waves called the real and the imaginary parts of the wavefunction. These components are equivalent. The term imaginary is a

FIGURE 6.1. *The wavefunction for a free particle with momentum p, which has wavelength, $\lambda = h/p$. A quantum mechanical wavefunction can have two parts, called real and imaginary. Both waves have the same wavelength. They are just shifted by one-fourth of a wavelength, which is the same as a 90° shift in the phase. These two components are separate from each other. They do not interfere either constructively or destructively. For a free particle with the well-defined value of the momentum, p, the wave function extends from positive infinity to negative infinity, $+\infty$ to $-\infty$.*

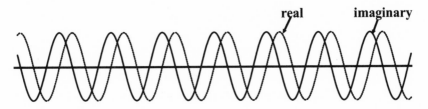

mathematical term. It does not imply that in some sense the imaginary component is less important than the part referred to as real. It is just jargon to identify the two components, although they do have differences in the way they are represented mathematically. The real and the imaginary components of the wavefunction have the same wavelength, but are shifted by one-fourth of the wavelength. That means that one wave is shifted in phase relative to the other by 90°. The two components of the wavefunction do not interfere with each other, either constructively or destructively, because in a mathematical sense and in essence they are perpendicular to each other.

A PARTICLE WITH WELL-DEFINED MOMENTUM IS SPREAD OVER ALL SPACE

The important feature of the wavefunction shown in Figure 6.1 is that it extends from positive infinity to negative infinity, that is, from $+\infty$ to $-\infty$. In Figure 6.1, only a small section of the wavefunction in a small region of space is shown because we cannot plot $+\infty$ to $-\infty$ on a finite piece of paper. The wave shown in the figure just keeps going to the right and to the left. It is uniform across all space. This means that for a quantum mechanical particle with a definite value of the momentum, p, we are equally likely to find the particle anywhere along the x axis, the horizontal axis in the graph. The vertical axis tells the probability amplitude of finding the particle somewhere. Both the real (dashed) and imaginary (solid) components oscillate positive and negative. Both have places where they are zero.

The fact that the wavefunction oscillates positive and negative doesn't matter. In explaining photon interference quantum mechanically following Figure 5.1, the Born interpretation of the wavefunction was introduced. In the Born interpretation, the absolute value squared of the wavefunction in a certain region of space gives the probability of finding a particle in that region of space. When the wavefunction is squared, it becomes only positive in the same way that $2^2 = 4$ and $(-2)^2 = 4$ because minus times a minus is a plus. Note that in Figure 6.1, whenever one wave is zero, the other wave is either at a positive or negative maximum. Where one wave is small, the other wave is big. When the wavefunction is analyzed mathematically, and as can be seen from the graph, at all locations the absolute value squared of the wavefunction is uniform along the x axis.

The absolute value squared of the wavefunction for a free particle is uniform along the x axis from $+\infty$ to $-\infty$. Therefore, the probability of finding the particle anywhere in space is uniform. The particle has equal probability of being found at $x = 10$, or $x = -1,000,000$, or anywhere else. Imagine that you are a tiny creature, frequently referred to as Maxwell's Demon. You are standing next to the particle-wave shown in Figure 6.1. You make a grab for the particle. There is some probability that you will find it in your hand. If you start over and do this again and again, depending on the size of your hand, you may eventually come up with the particle. Each time you need to start fresh in your attempt to grab the particle. If you move somewhere else along the wave and do the same thing, the chance that you will come up with the particle is no different. This is what it means to say that there is equal probability for finding the particle anywhere. There is no best spot for Maxwell's Demon to stand to try to grab the particle. All locations are equally good.

This picture of a free particle described by a wavefunction that represents equal probability of finding the particle anywhere doesn't go along very well with our classical concept of a particle. In Figure 2.5, we described a classical particle as having a particular momentum and position at a given time. In discussing the photoelectric effect (Figure 4.3), Einstein described light as photons, which are quanta of light. One photon "hits" one electron, and the electron flies out of the piece of metal. This description almost sounds like both the photon and the electron are particles in the classical mechanics sense of particles. However, in discussing the interference of photons in conjunction with Figure 5.1, it was necessary to use the Born interpretation and describe photons as probability amplitude waves, with half of the probability going into each leg of the interferometer. In Figure 6.1, the plot of a free particle wavefunction is completely delocalized, spread out over all space. The description is the same for a photon or an electron.

INTERFERENCE OF WAVES WITH DIFFERENT WAVELENGTHS

So what are photons and electrons and rocks and anything else? Are they particles or waves? To see that there is no contradiction in the quantum mechanical description of the nature of things, we need to discuss waves and the interference of waves further. In connection with Figures 3.2 and 3.3, we discussed that waves could interfere constructively to give a bigger wave or destructively to give a smaller wave or no wave at all. In the examples in Figures 3.2 and 3.3, the waves have the same wavelengths. When they added constructively (Figure 3.2), all of the positive peaks lined up with the positive peaks and the negative peaks lined up with the negative peaks to give increased amplitude. When the waves added destructively (Figure 3.3), the positive peaks lined up with the negative peaks and vice versa, to give cancellation. However, waves of different wavelengths can also interfere.

Figure 6.2 shows a plot of five waves with different wavelengths. The units of length do not matter. What is important is that the five waves have wavelengths, $\lambda = 1.2, 1.1, 1.0, 0.9,$ and 0.8. The phase of the waves are adjusted so that they all match at the point $x = 0$, where x is the horizontal axis. The waves match at $x = 0$ in the sense that each wave has a positive going peak at $x = 0$.

FIGURE 6.2. *Five waves are shown that have different wavelengths. The wavelengths are* λ *= 1.2, 1.1, 1.0, 0.9, and 0.8. The phases are adjusted so all of the peaks of the waves match at 0 on the horizontal axis. However, because the waves have different wavelengths, they do not match up at other positions, in contrast to Figure 3.2. Note that at a position of approximately 10 or −10, the dark gray wave has a positive peak, but the dashed light gray wave has a negative peak.*

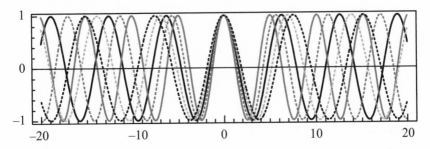

However, because the waves have different wavelengths, the peaks do not necessarily match at other points along the x axis. For example, at about x = 10 or −10, the dark gray wave has a maximum but the dashed light gray wave has a minimum. In addition, at approximately 10, one wave has a negative value and another has a positive value. At approximately x = 16 or −16, two waves have maxima, but another wave is at a minimum. The important point is that for waves with different wavelengths, at one point (x = 0 in the example), all of the waves can be matched, but in general, at other points, some of the waves will be positive and some of the waves will be negative.

Figure 6.3 shows the result of superimposing (adding up) the five waves in Figure 6.2. At x = 0 (horizontal axis) in Figure 6.2, all of the waves are exactly in phase. The superposition (adding the waves together) shown in Figure 6.3 yields a maximum. In Figure 6.2, the waves are all exactly in phase only at x = 0. Near x = 0, the difference in the wavelengths has not produced a large shift in the peaks of one wave relative to another, so the waves are still pretty much in phase. There is another set of maxima at about x = 6 and −6. However, these maxima are not as large as the one at x = 0, because the peaks of the waves are not all right on top of each other, as can be seen in Figure 6.2. Beyond x = ±10, the amplitude of the superposition is getting small. At any point, some waves are positive and some waves are negative, and they tend to destructively interfere. Because there are only five waves, the destructive interference is only partial.

Figure 6.4 shows the superposition of 250 waves with different wavelengths. The waves have equal-sized steps in wavelength in the range of wavelengths from 0 to 4. As for the five waves and their superposition shown in Figures 6.2 and 6.3, each wave has the same amplitude. The phases of the 250 waves are adjusted to match at x

FIGURE **6.3.** *The superposition of the five waves shown in Figure 6.2. At x = 0 (horizontal axis), all of the waves in figure 6.2 are in phase, so they add constructively. Near x = 0, the waves are still pretty much in phase, but the next set of maxima at about x = 6 and −6 are not as large as the maximum at x = 0. In the regions between 10 and 20 and −10 and −20, the difference in wavelengths makes some of the waves positive, where others are negative. There is significant cancellation.*

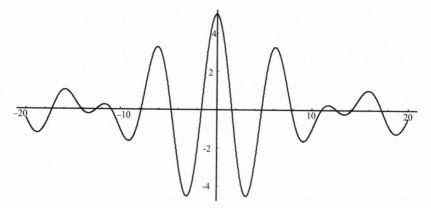

FIGURE 6.4. *The superimposition of 250 waves with equally spaced wavelengths spanning the wavelength range 0 to 4. Compared to Figure 6.3, which is the superposition of five waves, this superposition has a much larger peak at x = 0, the region of maximum constructive interference, and destructive interference reduces the other regions more. The amplitude of the superposition is dying out going toward + 20.*

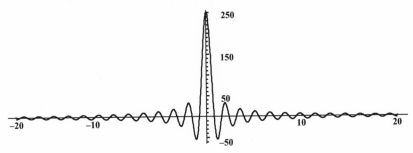

= 0 (x is the horizontal axis). Because there are many more waves over a wider range of wavelengths than in the superposition shown in Figure 6.3, the peak around x = 0 is much narrower, and the rest of the superposition dies out much more rapidly. The little oscillations come from the fact that all of the waves in the superposition have the same amplitude. If the amplitude of the wave at the middle of the spread of wavelengths has the biggest amplitude and the amplitudes of the other waves get smaller and smaller for wavelengths further and further from the center wavelength, it is possible to create a superposition that decays smoothly to zero without the set of deceasing amplitude oscillations. This type of superposition will be discussed below.

THE SUPERPOSITION PRINCIPLE

In Chapter 5, the interference experiment was analyzed in terms of the superposition of two photon translation states, T1 and T2. A photon in the interferometer is described as being in the 50/50 superposition state, T = T1 + T2. The idea of superposition is central to the quantum theoretical description of nature; it is called the Superposition Principle and assumes that "Whenever a system is in one state, it can always be considered to be partly in each of two or more states."

An original state can be regarded as a superposition of two or more states, as in the interference problem in which the translation state T of the photon was described as a superposition of T1 and T2. Conversely, two or more states can be superimposed to make a new state. It is this second statement that we will now use to understand the fundamental nature of particles. The fact that a photon can act like a particle in the photoelectric effect but act like a wave to give rise to the interference effect follows from the superposition principle and leads to the Heisenberg Uncertainty Principle.

Eigenstates

In connection with Figure 6.1, it was stated that a free particle with perfectly defined momentum p is a delocalized probability amplitude wave spread out over all space. If a particle exists in such a state, it is said to be in a momentum eigenstate. In discussing the interference problem, we called T1 and T2 pure states, but the correct name for them is eigenstates. Eigen is German for characteristic, so an eigenstate is a characteristic state. An eigenstate for a particular observable property, such as momentum, is a state with a perfectly defined value of that property. The momentum eigenstates of a free particle are delocalized over all space. One such eigenstate exists for each of the infinite number of possible values of the momentum. Each of these momentum eigenstates is associated with an exact value of momentum of the particle. The particle's location is uniform over all space because the wavefunction associated with the eigenstate is spread out over all space. However, the

Superposition Principle tells us that we can superimpose any number of momentum eigenstates to make a new state.

Superposition of Momentum Eigenstate Probability Amplitude Waves

To understand the nature of real particles, photons, electrons, and so on, we will superimpose a range of momentum eigenstate probability amplitude waves, such as the one shown in Figure 6.1. For each momentum p, the wave has a different wavelength, $\lambda = h/p$. In Figures 6.3 and 6.4, we saw that adding together waves of different wavelengths concentrated the amplitude of the wave in a particular region. As mentioned in both examples above, the amplitude of each wave in the superposition was the same. Now we will superimpose momentum probability amplitude waves with different amplitudes. There is one wave (a particular value of p) with the largest amplitude. The other waves with different wavelengths have amplitudes that decrease as the wavelength becomes greater than or less than the wavelength of the wave with the maximum amplitude. So, we have a distribution of wavelengths centered around the wavelength of the maximum amplitude wave. The wavelength with maximum amplitude is at the center of the distribution. By a distribution, we just mean that there is a range of wavelengths, in the same way that if you have a room full of people there will be a distribution of ages. There will be some people of average age, the center of the distribution, and some people older than the average and some people younger. Here, we have a wave at the center of the distribution with other waves having shorter wavelengths and still others having longer wavelengths.

Figure 6.5 illustrates a distribution of momentum probability amplitude waves. p_0 is the momentum of the wave at the center of the distribution of waves. It has a wavelength $\lambda = h/p_0$. It is the wave with the biggest amplitude, that is, the biggest probability of

FIGURE 6.5. *A plot of the probability of finding a particle in a particular momentum eigenstate with momentum p given that it is in a superposition of momentum probability amplitude waves. p_0 is the middle wave with the biggest amplitude in the distribution. Δp is a measure of the width of the distribution of eigenstates.*

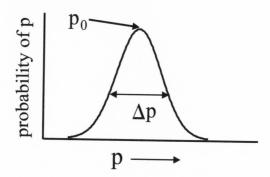

finding it in the distribution. As the momentum is increased or decreased (λ is smaller or bigger) away from p_0, the amount of a particular wave in the superposition (its probability) decreases. Δp is a measure of the width of the distribution. If Δp is large, then there is a large spread in p, and therefore, a large spread in wavelengths in the distribution. If Δp is small, the spread in wavelengths is small.

Momentum of a Free Particle in a Superposition State

What is the momentum of a free particle that is in a superposition of momentum eigenstates such as that shown in Figure 6.5? A superposition of momentum eigenstates just means that we add together (superimpose) a bunch of waves (probability amplitude waves) with each of the waves having a specific value of the momentum associated with it (an eigenstate). In any measurement of a property of a system, a particular value of that property will be measured. If we make a measurement of the momentum of a particle, we will measure a single value of the momentum. The nature of the disturbance that accompanies the measurement of an absolutely small object is to collapse the superposition state into a single eigenstate. Making a measurement changes a system by taking it from its initial superposition state to one particular eigenstate. This is what is meant by collapse.

In discussing the interference problem, we said that if we tried to find out if the photon was in the state T1 by placing a detector in leg 1 of the interferometer, we would destroy the superposition state necessary for the interference. The superposition state T would jump into either T1 or T2. Since the state T is a 50/50 superposition of T1 and T2, half the time a measurement will result in finding the system in T1 and half the time in T2. On any single measurement, it is impossible to know ahead of time which result will occur. Many measurements show that the superposition is 50/50 because half the time we find that the photon is in leg 1 of the apparatus (state T1) and half the time we find the system is in leg 2 of the apparatus (state T2).

The superposition of momentum eigenstates shown in Figure 6.5 is composed of a vast (infinite) number of states spread over a range of momenta characterized by the width of the distribution, Δp. Therefore, there is a wide range of momentum values that can be measured on any single measurement. If we make a single measurement, we will measure one of the many values. Let's say we make a measurement and find a momentum a little bigger than p_0. Call it p_1 because it is our first measurement. In the process of making the measurement, we made a nonnegligible disturbance of the system. It was changed from being in the superposition state to a single eigenstate with momentum p_1. So to make another measurement, we need to start over again and prepare the particle (the system) in the same way we did originally to get the same distribution of momenta. We make a second measurement. This time we measure a value that is quite a bit smaller than p_0. Call this value

p_2. We prepare the system again, and make another measurement. We measure p_3. Each time we make a measurement on an identically prepared system, we will measure a particular value of the momentum. In advance, we don't know what the value will be. If we make many, many measurements, we can plot the probability of measuring particular values of p. Such a plot gives the distribution like that shown in Figure 6.5. We can't say what value we will obtain on a single measurement. However, we do know something. It is unlikely that we will measure a value of p that is much much greater or much much smaller than p_0 because the distribution has very low amplitude in the wings (extremes) of the distribution. We are most likely to measure a value of p that is near p_0 because this is the portion of the distribution where the amplitude is large.

Momentum Not Perfectly Defined for a Particle in a Superposition of States

A particle in a superposition of momentum eigenstates, such as that shown in Figure 6.5, does not have a perfectly well-defined value of the momentum. In a single measurement, we cannot say what value of the momentum will be measured. We can say we are most likely to measure a value near p_0. With many measurements, we can determine the probability distribution. A classical particle, like the one illustrated in Figure 2.5, has a perfectly well-defined momentum. We can measure it without changing it. If it is a free particle, we can make many measurements of the momentum at different times, and we will always measure the same value of p. This is not the case for an absolutely small quantum particle in a momentum superposition state. We will measure a single value of p on a single measurement, but the act of making the measurement fundamentally changes the nature of the particle. The particle goes from being in a superposition state to being in an eigenstate (a single wave with a single value of the momentum). It goes from being in a state in which there is a probability distribution of momenta to a single value of the momentum that is observed. To recover the distribution, the particle needs to be prepared again.

WHERE IS A PARTICLE WHEN IT IS IN A MOMENTUM SUPERPOSITION STATE?

In connection with Figure 6.1, we said that a particle in a single momentum eigenstate is delocalized over all space. This doesn't go along well with the description of the photoelectric effect. Now the question is where is a particle that is in a momentum superposition state? We have already hinted at the answer with the discussions surrounding Figures 6.2 through 6.4. In Figures 6.3 and 6.4, we saw that a superposition of waves with different wavelengths produced a spatial distribution that was concentrated in a region of space. In Figure 6.3, the wavelengths went from 0.8 to 1.2, and the pattern was not as concentrated as the one in Figure 6.4, which was formed from wavelengths that went from 0 to 4. Figure 6.6 shows the spatial distribution associated with distribution of waves (momentum

FIGURE 6.6. *A plot of the probability of finding the particle at a location x given that it is in the superposition of momentum eigenstates shown in Figure 6.5. x_0 is the middle position with the greatest probability. Δx is a measure of the width of the spatial distribution.*

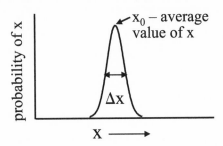

eigenstates) shown in Figure 6.5. There is a position where the value is maximum, which is also the average value. For larger and smaller values of x relative to x_0, the amplitudes (probabilities) become smaller.

What does the probability distribution of positions (x values) mean? A particle with the momentum probability distribution shown in Figure 6.5 gives rise to the spatial probability distribution shown in Figure 6.6. A single measurement of the position will measure a particular value of the position. Call it x_1. When the position measurement is made on the absolutely small quantum particle, it causes a nonnegligible disturbance that collapses the position probability distribution into a position eigenstate with a perfectly defined value of the position. To make another measurement, the system (a particle) must be prepared again in the identical manner so that it has the same momentum probability distribution and, therefore, the same spatial probability distribution. The second measurement of the particle's position will give a value, x_2, which in general will not be the same as x_1. If the system is prepared again and again, and many measurements of the position are made, the position probability distribution shown in Figure 6.6 will be mapped out. Δx is a measure of the width of the spatial distribution. The spatial distribution shown in Figure 6.6, which is determined by many measurements on identically prepared systems, tells the likelihood of measuring any particular position. A measurement is more likely to find the particle somewhere in the vicinity of x_0, but on any single measurement, it is impossible to say where the particle will be found. However, there is only a small probability of measuring a position that is far from x_0.

Wave Packets

A particle in a superposition of momentum eigenstates, such as that shown in Figure 6.5, is called a wave packet. Its momentum is more or less known depending on how big Δp is. Because the momentum is the mass times the velocity and the mass of a particle is known, we more or less know the particle's velocity. A bigger Δp (the wider the spread of momenta in the wave packet) results in the momentum being less well defined, which means that on any single measurement, one of a broader range of values of the momentum will be measured. The wave packet also has a spread in its position. The particle is not at a particular value of x as is a classical particle. There is a spread in positions given by the distribution like that in Figure 6.6, which can be quantified by the width Δx.

Spread in Momentum and Position

Figure 6.7 illustrates two wave packets. The top panels display a wave packet composed of a comparatively wide distribution of momentum eigenstates. The broad distribution of momentum eigenstates (large Δp) produces a spatial distribution that is relatively narrow (small Δx). The lower portion shows a wave packet composed of a relatively narrow distribution of momentum eigenstates (small Δp), which results in a relatively broad spatial distribution (large Δx).

The relationship between Δp and Δx illustrated in Figure 6.7 is general. A wave packet with a broad range of momenta (large uncertainty in momentum) will have a narrow spread of positions (small uncertainty in position). This relationship is produced by interference. A wave packet composed of a broad range of momentum eigenstates has a broad range of wavelengths because each momentum eigenstate has associated with it a probability amplitude wave with wavelength, $\lambda = h/p$. All of the probability amplitude waves in the packet can constructively interfere at some point in space. However, as shown in Figure 6.2, as the distance from this center point of constructive interference increases, destructive interference sets in. At any point far from the center, some waves will be positive

FIGURE 6.7. *The momentum (p) probability distributions and position (x) probability distributions for two wave packets. At the top, there is a large spread p (large Δp), which produces a small spread in x (small Δx). At the bottom, there is a small spread in p (small Δp), which gives rise to a large spread in x (large Δx).*

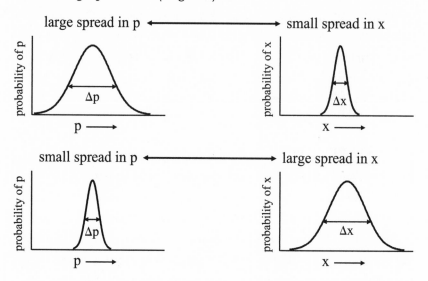

while other waves are negative, as can be seen in Figure 6.2. When the spread in wavelengths is large, the vast differences in the wavelengths cause the onset of destructive interference very close to the center point of maximum constructive interference, and the packet is narrow (large Δp, small Δx). When the spread in wavelengths is small, the wavelengths are not very different from one to another. Therefore, it is necessary to move far from the center point of perfect constructive interference before an equal number of waves will be positive and negative at a given point. In this case, Δp is small so Δx is large.

Because the idea of a spread in momentum and a related spread in position is so important, let's reprise the meaning of a spread. Everything is related to experiments. In a single experiment to measure the momentum of a particle, only one value can be measured. You have some instrument. It reads out a number. It can't tell you that the momentum is both 10 and 50 at the same time. In a single measurement, a single well-defined value will be measured. How do we get a single value when our wave packet has a spread in momenta? The wave packet is made up of a superposition of momentum eigenstates, that is, momentum probability amplitude waves with associated well-defined values of the momentum. When a measurement is made, the nonnegligible disturbance accompanying the measurement causes the system to "jump" from being in a superposition state to a particular eigenstate. The measurement gives the value of the momentum that goes with that eigenstate. Now, the measurement changed the system. To make another measurement, we need to start over and prepare the particle in the same way. By preparing it in the same way, a wave packet composed of the same superposition of momentum eigenstates will be generated. We now make the same measurement we did the first time. In general, we will measure a different value of the momentum because the wave packet is composed of many momentum waves each with a different observable value of the momentum associated with it. If we do this over and over again, each time preparing the wave packet in the same way and then making the measurement, we will get a spread in the measured values of the momentum. After a huge number of measurements, we might measure (neglecting units) the value 400 a thousand times, 390 eight hundred times, 410 eight hundred times, but 200 and 600 only twenty times. If we make a plot of all of these numbers we get a probability distribution like those shown for momentum on the left side of Figure 6.7. A probability distribution is an experimental determination of the composition of the wave packet. We now know how much (what is the probability) of each wave in the packet. The same description also applies to the position of our wave packet. On each measurement of the position of identically prepared wave packets, a single location for the particle will be found. After many measurements, a distribution of positions is determined like those illustrated on the right side of Figure 6.7.

THE HEISENBERG UNCERTAINTY PRINCIPLE

A very important point is that there is a relationship between the spread in momentum and the spread in position that is fundamental to the superposition state description of particles. When the spread in momentum (Δp) is large, there are many waves spread out along the x axis (see Figure 6.1) that combine to make the wave packet. These waves have different wavelengths (see Figure 6.2). When many waves with a wide range of wavelengths interfere, the region of constructive interference dies out very fast away from the maximum (see Figures 6.3 and 6.4). That means that the spread in position (Δx) is small. If there is only a small range of the momentum waves that make up the wave pack (Δp small), then the spatial region of constructive interference dies out slowly away from the maximum in the position distribution (see Figure 6.7). Therefore, the spread or uncertainty in position Δx is large. All of this occurs because of the probability amplitude wave nature of the wave functions that describe the momentum eigenstates. A wave packet is located, more or less, in the region of constructive interference, and there is little probability of finding the particle in the regions of substantial destructive interference.

The formal relationship between the spread in the momentum and the spread in the position, that is, between Δp and Δx, is called the Heisenberg Uncertainty Principle. Werner Karl Heisenberg (1901–1976) won the Nobel Prize in Physics in 1932 "for the creation of quantum mechanics, the application of which has, inter alia, led to the discovery of the allotropic forms of hydrogen." The Heisenberg Uncertainty Principle is stated through a simple mathematical relationship, $\Delta x \Delta p \geq h/4\pi$, where h is Planck's constant and the Δx and Δp define the widths of the distributions for position and momentum, as shown in Figure 6.7. ($\geq$ means greater than or equal to.) Whether the equal sign ($=$) or the greater than sign ($>$) applies depends on the shapes of the probability distributions. The equal sign applies for a shape called a Gaussian after the great mathematician, Karl Friedrich Gauss (1777–1855). The curves shown in Figures 6.5 through 6.7 are Gaussians. A Gaussian is the standard "bell-shaped curve" that describes the distributions of things, such as test scores, for a properly designed test with a large enough number of people taking the test. Gaussian-shaped curves show up in many places in the physical sciences. The greater than sign applies to other shapes. For any shape, which is made up from a specific distribution of waves, it is possible to determine what the product $\Delta x \Delta p$ will be, but it will always be $> h/4\pi$ unless the shape is a Gaussian.

To understand the nature of the Uncertainty Principle, it is sufficient to consider Gaussian shapes like those in Figure 6.7. Then, $\Delta x \Delta p = h/4\pi$. The equation shows what it is possible to know simultaneously about the position and momentum of a particle. $h/4\pi$ is a constant. Therefore, $\Delta x \Delta p$ equals a constant. So if the uncertainty in the momentum, Δp, is large, then the uncertainty in the position, Δx, must be small, so that the product is $h/4\pi$. On the other hand, if Δp is small, then Δx is large. The connection between Δp and Δx is illustrated in Figure 6.7. The uncertainty principle says that you can know something about the momentum of a particle and something about the position of a particle, but you can't know both the position and the momentum exactly at the same time. This uncertainty in the simultaneous knowledge of the position and the momentum is in sharp contrast to classical mechanics. It is fundamental to classical mechanics theory, as illustrated in Figure 2.5, that the position and the momentum of a particle can be precisely known (measured) simultaneously. Quantum theory states that it is impossible to know both the position and the momentum precisely at the same time. We can know both within some uncertainties, Δx and Δp.

Examining the Uncertainty Principle relationship, $\Delta x \Delta p = h/4\pi$, consider what happens as we make Δp smaller and smaller. As Δp becomes smaller and smaller, Δx grows. Dividing both sides of the equation by Δp gives $\Delta x = \dfrac{h}{4\pi\Delta p}$. As Δp becomes smaller, we are dividing by a smaller and smaller number, so Δx grows. As Δp gets closer and closer to zero, Δx gets closer and closer to infinity. In the limit that Δp goes to zero, Δx goes to infinity. This limit has an important meaning. If Δp is zero, the momentum is known precisely, but the position is totally unknown. With $\Delta x = \infty$, the particle can be found anywhere with equal probability. This result is in accord with the discussion surrounding Figure 6.1, which shows the wavefunction for a momentum eigenstate. When a particle is in a momentum eigenstate, it has a perfectly well-defined value of its momentum. However, its probability amplitude function, which describes the probability of finding the particle in some region of space, is spread out (delocalized) over all space. The probability of finding the particle anywhere is uniform; $\Delta x = \infty$. This is in contrast to the wave packets shown in Figure 6.7, where a superposition of momentum eigenstates produces a state in which there is no longer a perfectly well-defined momentum, but there is some knowledge of the position. We know the position and momentum within some ranges of uncertainty.

If we rearrange the uncertainty relation to give $\Delta p = \dfrac{h}{4\pi\Delta x}$, we see that in the limit that Δx goes to zero (perfect knowledge of the position), Δp goes to infinity. If we know the position perfectly, the momentum can have any value. A wave packet composed of all of the momentum eigenstates ($\Delta p = \infty$) has a perfectly well-defined value of the position. It is possible to know p precisely but with no knowledge of x; it is possible to know x precisely, but with no knowledge of p. This is called complementarity. You can know x or p but not both at the same time. In classical mechanics, you can know x *and* p. In quantum mechanics, you can know x *or* p. Generally for quantum particles, absolutely small particles, you know something about p and something about x, but you can't know both precisely simultaneously.

7

Photons, Electrons, and Baseballs

PHOTONS, ELECTRONS, AND BASEBALLS are all described with quantum theory in the same way, but quantum theory isn't necessary to describe baseballs. Baseballs act as classical particles and behave in a manner that is accurately described by classical mechanics. If photons, electrons, and baseballs all have the same quantum mechanical description, why do only baseballs act as classical particles? The answer is that baseballs are large in the absolute sense. Here, we will see why photons and electrons need a quantum theory description but baseballs do not. Real physical situations are discussed that bring out both the wavelike and particlelike nature of quantum particles, that is, absolutely small particles.

WAVES OR PARTICLES?

When a particle is in a superposition state, a wave packet, we have some knowledge of its position and some knowledge of its momentum. So are photons, electrons, and so on waves or particles? The answer is that they are wave packets. Whether they seem to be particles or seem to be waves depends on what experiment you do, that is, what question you ask. In the photoelectric effect, the photons act like particles. One photon hits one electron, and kicks it out of the metal (Figure 4.3). The photon is a wave packet formed from a spread of momentum eigenstates. The width of the spread, Δp, results in a relatively well-defined position, that is, a relatively small Δx. Thus, the photon wave packet is more or less well defined in position, so that it can act like a particle of light in the photoelectric effect. In the interference experiment (Figure 5.1), the photons act like waves. This may not be surprising because the wave packet is, in fact, a superposition of waves, but not waves in the normal classical sense, but rather probability amplitude waves. In the discussion of the interference phenomenon, the photon wave is discussed as if it is a single probability amplitude wave. Now it is clear that it is actually a wave packet, composed of a superposition of waves. When it hits the beam splitter, it becomes a superposition of the two translation states, T1 and T2. The probability amplitude waves in T1 interfere with their corresponding waves in T2, to produce the interference pattern as discussed above.

DIFFRACTION OF LIGHT

So a photon acts like a particle in the photoelectric effect, but it can also act like a wave. An experiment that clearly shows the wavelike properties of photons is diffraction of light from a diffraction grating. It is possible to see diffraction using a music compact disk (CD) and a bright light or sunlight. The CD has very fine grooves on its surface. These are the tracks on which the information is stored. As explained below, when white light from the sun or a light bulb falls on the CD, the grooves diffract the light, sending different colors in particular directions. Different parts of the CD make a range of angles relative to your eye, which causes the appearance of different colors emanating from different parts of the CD.

Diffraction from a grating is used in many optical instruments, called spectrometers. These instruments separate the different colors of incoming light so that the colors can be analyzed individually. A recording of the colors of light from a particular source is called a spectrum. For example, stars emit different colors of light depending on their temperature. Taking a spectrum of the light from a star can provide a great deal of information about the star. Star light traveling through space will encounter molecules in space. As discussed in Chapter 8 and subsequent chapters, different molecules absorb different colors of light. Star light traveling to Earth has some of its colors partially absorbed by the molecules in space. Astronomers mount spectrometers on telescopes and take spectra to determine what types of molecules are in space between a particular star and the Earth.

Figure 7.1 shows the geometry for light diffraction from a grating. The incoming light is at an angle α (Greek letter alpha) relative

FIGURE 7.1. *Geometry of light diffraction from a grating. The grating is composed of a reflective surface, usually silver or gold, with very fine parallel grooves in it. The grating is shown here from the side. The grooves run into the page. The grooves have a very uniform spacing, d; α is the angle of the incoming light. The outgoing angles β depend on the color. Therefore, the colors are separated by diffraction.*

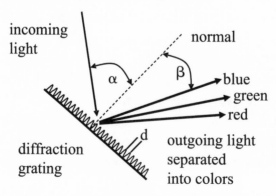

to the normal to the grating. The normal is the direction perpendicular to the surface of the grating. The grating is shown from the
side. The surface of the grating, which looks like a flat mirror to the
eye, has a dense set of parallel grooves in it. These grooves are called
lines. The spacing between the lines is labeled d in Figure 7.2. The
spacing is about the wavelength of light, approximately one ten millionth of a meter. The grooves are highly reflective. They are usually
gold or silver. For incoming light that has a range of colors, the
outgoing light will be separated by color such that each color goes
in a unique direction. The separation into colors going in different
directions is illustrated in Figure 7.1. The angle between the normal
to the grating and a particular color is labeled β (Greek letter beta)
in the figure. β is shown for the blue color. β is bigger for green
and still bigger for red.

Diffraction of Light Shows Wave Character of Photons

The diffraction of light from a grating shows the wave nature of
photon wave packets. To see that diffraction brings out the wave
character of photons, we need to look at how diffraction works in
terms of constructive and destructive interference of waves. Figure
7.2 shows incoming photon wave packets as a beam of light impinging on the diffraction grating. The light travels different distances
before it strikes various parts of the grating. The light that reaches
the upper left portion of the grating will travel a shorter distance
than the light that hits the bottom right portion of the grating. The
wave packet is composed of many colors, that is, many waves with
wavelengths, λ. The different colors of light will come off of the
grating in all directions. Here is the tricky part. The wave packets
are more or less localized, but they are composed of different colors,
each of which is a delocalized probability amplitude wave (see Figures 6.1, 6.2, 6.4, and 6.7). The more or less localized wave packet
is formed by the interference of many different color waves (differ-

ent λs, which correspond to different momenta, p). Consider one particular color, red, that comprises part of the wave packet. If the wave hits only one line in the grating, it would reflect off in many directions because of the shape of the groove. It would leave the single groove as a superposition state composed of probability amplitude waves propagating in many directions. In the interferometer (Figure 5.1), the incoming wave packet became a superposition state that had probability amplitude propagating in two directions. Here, after hitting a single line, the superposition would be heading in many directions.

The important feature of a grating is that the incoming wave hits many lines in the grating. For a particular color, red shown in Figure 7.2, there is a single direction in which the waves will add up constructively. In the figure, for the direction in which the red waves are propagating, all of the peaks and troughs of the waves add in phase even though they reflect from different places. (The wavelength has been exaggerated relative to the line spacing, d, to make it easy to see the alignment of the waves.) The in phase addition of many waves leaving the grating makes a very large outgoing wave. In all other directions, the red waves will add destructively because the peaks and null do not line up.

Diffraction from a grating causes the waves of a given wavelength (a particular color) to add constructively in one direction. The intensity associated with a probability amplitude light wave is proportional the square of the wave amplitude. Therefore, in the direction of constructive interference for a particular color, red in the example, the intensity of light is large. In other directions, red light will experience destructive interference, because the wavelength is such that the differences in the distances from each groove don't equal an integral number of wavelengths. For another color, say blue, there is a different direction in which light coming off of all of the grooves will add constructively (see Figure 7.1). Therefore, the blue light component of the incoming photon wave packets will

FIGURE 7.2. *Incoming photon wave packets are diffracted from a grating. The different colors reflect off of the grooves. For a particular color, there is a direction in which the waves corresponding to that color constructively interfere. They add to make a large amplitude wave, so the color looks very bright in that particular direction.*

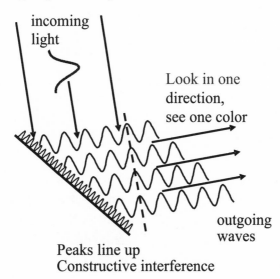

incoming
light

Look in one
direction,
see one color

outgoing
waves

Peaks line up
Constructive interference

leave the grating as a large amplitude wave in its own direction, and in this direction the intensity of the blue component of the incoming light will appear bright.

Electrons Act Like Bullets in a Cathode Ray Tube

Diffraction of light from a grating brings out the wave property of the photon wave packets, while the photoelectric effect demonstrates the more localized particle like properties of a photon wave packet. In the discussion of the de Broglie wavelength, which relates the momentum to a wavelength through the relation $p = h/\lambda$, it was mentioned that the descriptions of electrons and other types of "particles" are the same as the description of photons. Both photons and electrons are described in terms of probability amplitude waves. Both are more or less localized wave packets (see Figure 6.7). For an electron that is a free particle (no forces are acting on it) the wave packet is a superposition of the free particle momentum eigenstates. The uncertainty in the electron's position, Δx, is determined by the uncertainty (spread) in the momentum, Δp, through the Heisenberg's uncertainty relationship, $\Delta x \Delta p \geq h/4\pi$. The equality sign holds for Gaussian wave packets, which are the shapes shown in Figure 6.7.

To illustrate both the particle nature and wave nature of electrons, two examples are discussed: how a cathode ray tube (CRT) works and low energy electron diffraction from a crystal surface. Cathode ray tubes used to be ubiquitous. They are the devices that produce the pictures in the original televisions and computer monitors. CRTs are the large, boxy TVs and monitors that are rapidly being replaced by other devices, such as liquid crystal displays (LCDs). (There are actually several technologies used to make large-screen thin TVs, but thin computer monitors are all LCDs.)

Figure 7.3 is a schematic diagram of a CRT. Inside the CRT is a vacuum in which electrons can move without colliding with air molecules. The process that produces the picture begins with the filament, a piece of wire (left side of figure). An electrical current is passed through the filament, which causes it to get very hot like the filament in a conventional light bulb, the element in an electric stove, or the element in an electric space heater. The heat from the filament heats the cathode until it is also very hot. The cathode is a piece of metal that is connected to a negative voltage, like the negative end of a battery, but at a much higher voltage. The cathode becomes so hot that electrons boil off. Heat is a form of energy. The electrons are held in the metal by a binding energy that depends on the type of metal. When the metal is hot enough, the thermal energy can overcome the electron binding energy and some electrons will leave the metal. In the photoelectric effect, a photon provided

FIGURE 7.3. *A schematic of a cathode ray tube (CRT). The hot filament heats the cathode, which "boils" off electrons. The positively charged acceleration grid accelerates the negatively charged electrons. Voltages applied to the control grids steer the electrons to particular points on the screen. The screen is covered with tiny adjacent red, green, and blue spots that glow with their particular color when hit with electrons. By rapidly scanning the electron beam to hit the appropriate colors in a given spot on the screen the image is made.*

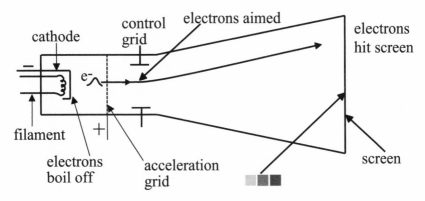

the energy to overcome the electron binding to the metal. In a CRT, heat provides the energy to eject the electrons from the metal. The electrons that leave are replaced by the connection to the negative power supply that puts electrons back into the cathode, so the process can go on continuously. Electrons are negatively charged, and because the cathode is at a negative voltage, the electrons are repelled from the cathode. Therefore, the electrons fly away from the cathode. The movement of the electrons away from the cathode is helped by the positively charged acceleration grid (see Figure 7.3). Because the acceleration grid is connected to a positive power supply, the negatively charged electrons are attracted to it. Like charges repel. Unlike charges attract. The acceleration grid is a mesh of wires that is mostly holes. When the electrons reach the grid, most of them fly right through it and keep going, moving very fast.

The electrons then pass between the control grids (see Figure 7.3), which control the direction the electrons go. There is one pair of control grids for the vertical direction (shown in the figure) and an equivalent set for the horizontal direction (not shown). Consider the vertical direction. If a positive voltage is applied to the top control grid and a negative voltage is applied to the bottom control grid, the electrons will be deflected up, as shown in Figure 7.3, because the negatively charged electrons will be attracted toward the positive top grid and repelled from the negative bottom grid. If the polarity of the voltages on the two grids is reversed, the electrons will be deflected downward. If large voltages are applied to the grids, the electrons will be deflected a lot. If small voltages are applied, the electrons will be deflected a small amount. If no voltages are applied, the electrons will go straight ahead. The same thing happens by applying voltages to the horizontal control grids. Once the electrons pass by the control grids, they continue in a straight line. In this manner, the electrons can be aimed just like bullets. This part of the CRT is referred to as an electron gun. Electron guns are used in many scientific devices such as electron microscopes and the device discussed below. So even when there are no more CRTs used as TVs and computer monitors, the basic device described here will still be important.

Because there is no air and gravity is a very weak force, the electrons travel basically as free particles until they hit the screen shown on the right side of Figure 7.3. On the screen are very small and very closely spaced patches of chromophores. Chromophores are chemical species that emit light when excited, that is, when sufficient energy is imparted to them. In this case, the chromophores are excited when the electrons hit them. In each very small region of the screen, there are three chromophores, one red, one green, and one blue. The electron beam can be aimed to hit a particular spot with great accuracy. If at a given location the red chromophore is hit, the screen lights up for an instant with a tiny red dot of light. If the green chromophore is hit, there is a green dot of light, and if the blue chromophore is hit, a blue dot of light is generated,

The electronics that produce the voltages on the control grids sweep the electron beam across the screen horizontally, then move the beam down, and sweep horizontally again. This is continued until the entire screen is swept. The beam returns to the top, and the sweep is repeated. As the beam sweeps, it is directed to hit red, green, or blue chromophores. The patches of three chromophores are so close together horizontally and vertically that your eye cannot distinguish them as individual dots. The beam can also be turned off so if no chromophore is hit you get a black spot. The combination of the three colors is sufficient to make any color. The image that we see on a CRT computer screen or a TV is formed by controlling which colors are hit and which spots are not hit at all. The electron beam is moved across the screen so fast that our eyes cannot tell that what we see is actually a very rapid succession of still pictures.

In the description of the CRT, the electrons acted very much like our general conception of particles. They could be aimed by the electron gun to hit very specific spots on the screen. This does not sound very much like a wave. However, it is certainly within our description of more or less localized wave packets. As long as Δx of the electron wave packets is small compared to the size of the chromophore spots (pixels), the fact that the wave packet is delocalized over a distance scale Δx doesn't matter. The colored pixels are small, but not small on the "absolutely small" distance scale. They are smaller than the eye can see without a microscope, but that is still large compared to the length scales that are encountered in atomic and molecular systems. Therefore, wave packets with a reasonably small Δp can still have an uncertainty in position that is very small compared to the pixel size. For a particle, such as an electron, $p = mV$, where m is the mass of the electron and V is the velocity of the electron. The mass is well defined. The uncertainty in p comes from an uncertainty in the velocity. So what Δp means is that the velocity is not perfectly well defined. Measurements of the velocity on identically prepared electron wave packets will not give the same value from one measurement to the next. The uncertainty in the velocity yields an uncertainty in the momentum, Δp, which through the Uncertainty Principle, $\Delta x \Delta p \geq h/4\pi$, gives the uncertainty in x, Δx. The important point is that Δx can be significant on the distance scale of atoms and molecules but very small compared to the distance scale of the macroscopic colored pixels on the CRT screen. In such situations, the wave nature of a wave packet is not manifested, and the wave packet behaves like a classical particle.

Electrons Act Like Waves in Electron Diffraction

Electron wave packets also display their wave properties, as illustrated in Figure 7.4. In this experiment, a beam of electrons generated with an electron gun, like that described above, is aimed at a crystal surface rather than a TV screen. The electrons are not of sufficiently high energy to penetrate the crystal. The surface of the crystal is composed of rows of atoms called a lattice or crystal lattice. The rows of atoms are spaced a few angstroms apart; one angstrom is 1×10^{-10} m, or one ten billionth of a meter. On the atomic distance scale, the unit of angstroms comes up often. It is given a special symbol, Å. The spacing is determined by the size of the atoms. The rows of atoms act as the grooves in a diffraction grating, but they are much more closely spaced. The wavelength of the electrons is on the same distance scale as the lattice spacing (row spacing). The wavelength is given by the de Broglie relation $\lambda = h/p$. p $= mV$. The mass of the electron is 9.1×10^{-31} kg (kilograms). Then for a velocity of 7.3×10^5 m/s (730,000 meters per second, about 1.6 million miles per hour), $\lambda = 10$ Å. This velocity is very easily obtained with a simple electron gun.

FIGURE 7.4. *A schematic of low-energy electron diffraction (LEED) from the surface of a crystal. An incoming beam of electrons, with low enough energy not to penetrate the crystal, strikes the surface. The lines of atoms act like the grooves of the diffraction grating in Figure 7.1. They diffract the incoming electron waves.*

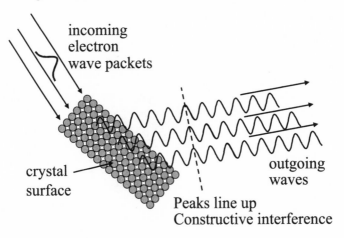

The electron probability amplitude waves diffract from the crystal surface in a manner akin to the photons diffracting from the ruled grating discussed above. However, in the ruled grating, there is a single separation, d, because the grooves all run parallel to each other in a single direction. The lattice at a crystal surface is two-dimensional. As can be seen in Figure 7.5, there are many directions in which there are parallel rows of atoms. The solid lines show rows of atoms running in different directions. The dashed line parallel to each solid line shows that there are parallel rows of atoms for each of the directions indicated by the solid lines. For the rows of atoms running in different directions, the spacing between the atoms (the diffraction groove spacing) is different. The difference in spacing can be seen in Figure 7.5. Look at the separation between a pair of solid and dashed lines. Each pair has a different separation, which is the groove spacing.

FIGURE 7.5. *The lattice from Figure 7.4, with examples of different rows of atoms shown by the lines. For each line passing through the centers of atoms in a row, it is possible to draw more lines that are parallel to the initial line and that also pass through the centers of atoms. The spacings between these distinct parallel rows are different. Each set of rows causes diffraction in a different direction.*

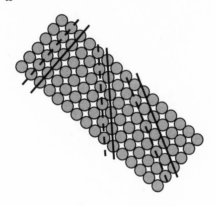

Because there are many different atom-to-atom spacings with the "grooves" running in different directions, the electrons' waves will be diffracted in many different directions. Figure 7.6 is an example of low-energy electron diffraction from a crystal surface. The black circle in the center is a piece of metal called a beam stop. It is supported by another piece of metal that appears as the vertical bar below the beam stop in the picture. The beam stop prevents the portion of the electron beam that is reflected from the crystal from hitting the detector. The brighter and dimmer white spots are produced by the diffracted electrons hitting the detector. From the location of the spots, the spacing and arrangement of the atoms can be determined. Electron diffraction from crystal surfaces is an important tool in the science of understanding the nature of surfaces. The electron diffraction pattern demonstrates conclusively that electrons can behave as waves, just like photons.

FIGURE 7.6. *Experimental data showing diffraction of electrons from the surface of a crystal. The various light spots are the electron diffraction spots. There are many spots because the diffraction occurs from the many different parallel rows of atoms (see Figure 7.5).*

ELECTRONS AND PHOTONS ARE PARTICLES AND WAVES, BUT BASEBALLS ONLY PARTICLES

Electrons act as particles in a CRT just as photons act as particles in the photoelectric effect. Electrons act as waves in low-energy electron diffraction just as photons act as waves when they diffract from a diffraction grating. In reality, photons, electrons, and all particles are actually wave packets that are more or less localized. Wave packets can display their wavelike properties or their particlelike properties depending on the circumstances.

If photons and electrons can show both wavelike and particlelike properties, why don't baseballs? To see the reason that baseballs act like particles in the classical mechanics sense, we need to look at the wavelengths associated with particles versus their size.

First consider an electron in an atom such as hydrogen. We are going to talk about the quantum description of the hydrogen atom and other atoms in Chapters 10 and 11, but for now, we will only use a very simple qualitative discussion of the wavelike characteristics of the hydrogen atom. The de Broglie relation tells us that the wavelength λ = h/p. The momentum is p = mV, the mass times the velocity. The mass of an electron is m_e = 9.1×10^{-31} kg. In an atom, the typical velocity of an electron is V = 5.0×10^6 m/s. Then the de Broglie wavelength is

$$\lambda = \frac{h}{p} = \frac{6.6 \times 10^{-34} \text{ J-s}}{(9.1 \times 10^{-31} \text{ kg})(5.0 \times 10^6 \text{ m/s})} = 1.5 \times 10^{-10} \text{ m} = 1.5 \text{Å}.$$

Note that 1.5 Å is approximately the size of an atom. Therefore, the wavelength of an electron in an atom is about the size of the atom. The wave properties of electrons will be very important when electrons are in systems that are very small like atoms.

What about a baseball? According to the rules of Major League Baseball, a baseball must weigh between 142 g and 149 g. We will take the mass to be 145 g = 0.145 kg. A pretty fast pitch goes 90 mph. 90 mph = 40 m/s. The momentum of a fast ball is p = 0.145 kg × 40 m/s = 5.8 kg-m/s. The de Broglie wavelength of the fast ball is:

$$\lambda = \frac{h}{p} = \frac{6.6 \times 10^{-34} \text{ J-s}}{(0.145 \text{ kg})(40 \text{ m/s})} = 1.1 \times 10^{-34} \text{ m} = 1.1 \times 10^{-24} \text{ Å}.$$

This is an unbelievably small number. The size of one atom is about 1 Å. The size of the nucleus of an atom is about 10^{-5} Å. Then the wavelength of a baseball is 0.0000000000000000001 of the size of one atomic nucleus. This wavelength is small beyond small. It is so small that it will never be manifested in any measurement. There can never be a diffraction grating with line spacing small enough to show diffraction for a wavelength that is a ten millionth of a trillionth of the size of an atomic nucleus. Because this wavelength is so small we never have to worry that a baseball will diffract from a baseball bat. It will always act like a classical particle. Objects that are large in the absolute sense have the property that the wavelengths associated with them are completely negligible compared to their size. Therefore, large particles only manifest their particle nature; they never manifest their wave nature. In contrast, particles that are small in the absolute sense have de Broglie wavelengths that are similar to their size. Such absolutely small particles will act like waves or act like particles depending on the situation. They are wave packets. In the manner that we have discussed, they are both waves and particles.

8

Quantum Racquetball and the Color of Fruit

IN THE PREVIOUS CHAPTERS, the fundamental concepts of quantum theory were introduced and explained. The examples given, however, looked only at the behavior of free particles. It was shown that electrons could behave as particles in the discussion of how a CRT works, but they behaved as waves in the description of electron diffraction from crystal surfaces. A free particle can have any energy. Its energy, which is kinetic, is determined by its mass and its velocity. As the velocity increases, the energy increases. A tiny increase in velocity produces a tiny increase in energy. A large increase in velocity produces a large increase in energy. The steps in energy can be any size; they are continuous. Bound electrons were discussed briefly in connection with the photoelectric effect. It was pointed out that if the energy of the incoming photon is insufficient to overcome the binding of electrons in the metal, no electrons will be ejected from the metal. Electrons bound to nuclei of atoms are responsible for the properties of atoms and molecules. It was also mentioned that Planck explained black body radiation, which will be discussed in detail later, by postulating that bound electron energies can only change in discreet steps. To understand the properties of the atomic and molecular matter that surrounds us in everyday life, it is necessary to treat bound electrons with quantum theory.

The essential feature of electrons bound to an atom or molecule is that their energy states are discreet. We say that the energies an electron can have are quantized, that is, an electron bound to an atom or molecule can only have certain energies. The energy goes in steps, and the steps are certain discreet sizes. The energy states are like a staircase. You can stand on one stair, or you can stand on the next higher stair. You cannot stand halfway between two stairs. These discreet or quantized energies are frequently called energy levels. Unlike a staircase, the energy levels are not generally equally spaced.

An important area of modern quantum theory research is the calculation of the electronic quantum states of molecules. This field is called quantum chemistry. Such calculations yield the quantized energies of electrons in molecules (energy levels), and they also calculate the structures of molecules. Molecular structure calculations give the distances between atoms and the positions of all atoms in a molecule within limits set by the uncertainty principle. Thus, quantum mechanical calculations are able to determine the size and shapes of molecules. Such calculations are important for understanding the basic principles of the bonding of atoms to form molecules and to design new molecules. As quantum theory continues to develop and the ability to solve complex math problems on increasingly fast and sophisticated computers continues to advance, larger and larger molecules can be investigated using quantum chemistry. One area of great importance is the quantum theory design of pharmaceuticals. Molecules can be designed to have the right size and shape to "fit" into a particular location in a protein or enzyme.

Quantum chemistry is mathematically very intense. Even the quantum mechanical calculation of the simplest atom, the hydrogen atom, is mathematically complex. The hydrogen atom consists of a single electron bound to a single proton. The proton, which is the nucleus of a hydrogen atom, is positively charged and the electron is negatively charged. It is the attraction of the negative electron to the positive proton that holds the hydrogen atom together. The detailed calculation of the energy levels of the hydrogen atom will not be presented, but in the next chapters, the results of the calculations are discussed in some detail. The calculations give the hydrogen atom energy levels and the hydrogen atom wavefunctions. The wavefunctions, that is, the probability amplitude waves for the hydrogen atom, are the starting point for understanding all atoms and molecules. Atoms and molecules are complicated because they are absolutely small three-dimensional systems, and it is necessary to deal with how protons and electrons interact with each other.

THE PARTICLE IN A BOX—CLASSICAL

There is a very simple related problem called the particle in a box. We can solve this problem without complicated math. The solutions to the particle in a box problem make it possible to illustrate many of the important properties of bound electrons, such as quantized energy levels and the wavelike nature of electrons in bound states. Before analyzing the nature of an electron in an atomic-sized one-dimensional box, the classical problem of an ideal one-dimensional racquetball court will be discussed so that the differences between a classical (big) system and a quantum mechanical (absolutely small) system can be brought out.

Figure 8.1 is an illustration of a perfect "box." It is one dimensional. The walls are taken to be infinitely high, infinitely massive, and completely impenetrable. There is no air resistance inside the box. In the figure, the inside of the box is labeled as $Q = 0$, and outside of the box, $Q = \infty$. Earlier we said that a free particle is a particle with no forces acting on it. For a force to be exerted on a

FIGURE 8.1. *A perfect one-dimensional box. The walls are infinitely high, infinitely thick, infinitely massive, and completely impenetrable. There is no air resistance in the box. In the box, Q, the potential energy is zero, and outside the box, it is infinite. The box has length, L.*

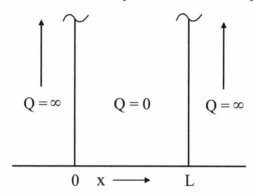

particle, there must be something for the particle to interact with. For example, a negatively charged particle like an electron can interact with a positively charged proton. The attractive interaction between the oppositely charged particles will exert a force on the electron. In the description of electron steering in a CRT (see Figure 7.3), an electric field exerted a force on the electrons, causing them to change direction. A measure of the interaction of a particle with some influence on it, such as an electric field, is called a potential and has units of energy. In the figure, the potential is labeled Q. Inside the box, $Q = 0$, just as in a free particle. This means the particle is not interacting with anything inside the box. There are no electric fields or air resistance. However, outside the box, $Q = \infty$. An infinite potential for the particle means the particle would have to have infinite energy to be located in the regions outside the box. $Q = \infty$ is just the formal way of stating that the walls are perfect. There is no possibility of the particle penetrating the walls or going over the top of the walls no matter how much energy it has. Therefore, if a particle is put inside the box, it is going to stay inside the box. There is no way for it to escape. In this sense, the particle is bound inside the box. It can be in the space of length L, but nowhere else.

Figure 8.2 shows a racquetball bouncing off of the walls in a perfect one-dimensional classical (big) racquetball court. As discussed, the walls are perfect, and there is no air resistance. In addition, the ball is perfect, that is, it is perfectly elastic. When a ball hits a wall, it compresses like a spring and springs back, which causes the ball to bounce off the wall. A real ball is not perfectly elastic. When the ball compresses, not all of the energy that went into compressing the ball goes back into pushing the ball off of the wall. Some of the energy that went into compressing the ball goes into heating the ball. However, here we will take the ball to be perfectly elastic. All of the kinetic energy of the ball that compresses it when it hits the wall goes into pushing the ball off of the wall.

FIGURE 8.2. *A ball in a perfect one-dimensional racquetball court. There is no air resistance, and the ball is perfect. When the ball strikes the wall at L, it bounces off, hits the wall at 0, and keeps bouncing back and forth. Because the court is perfect, the ball is perfect, and there is no air resistance; once the ball starts bouncing, it keeps bouncing back and forth indefinitely.*

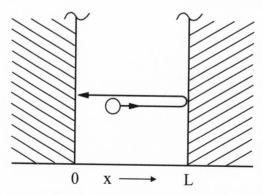

Therefore, the speed of the ball just before it hits the wall is equal to the speed of the ball after it bounces off.

In this perfect racquetball court, the ball bounces off of the walls without losing any energy and there is no air resistance or gravity. Therefore, the ball will bounce back and forth between the walls indefinitely. It will hit the wall at position L, bounce off, and then hit the wall at position 0, bounce off, and just continue bouncing back and forth. Because the potential inside the box is zero (see Figure 8.1), there are no forces acting on the ball. Therefore, its energy is purely kinetic energy, $E_k = \frac{1}{2}mV^2$; m is the mass of the ball and V is its velocity. If the ball is hit a little bit harder, it will go a little bit faster, that is, V will be a little bit bigger. E_k will be a little bigger. If the ball is hit somewhat more gently, the ball's velocity will be a bit smaller, and E_k will be a little smaller. In this perfect racquetball game, the energy can vary continuously. E_k can go up or down by any amount, the amount depending only on how hard you hit the ball.

Another important feature of classical racquetball is that it is possible to stop the ball and put it on the floor. In this situation, the ball has no velocity, V = 0. If V = 0, then E_k = 0. If V = 0, then the momentum is zero because p = mV. So we know the momentum exactly. If the ball is placed on the floor with V = 0, then the position is known. If the position is called x (see Figure 8.2), x can take on values between 0 and L. It cannot have any other values because the ball is inside the court (the box) and can't get out because of the perfect walls. The ball can be placed at a certain position x on the floor of the court. Therefore, its position is known exactly. This is a characteristic of a macroscopic racquetball court, even a perfect one. It is a classical system, and it is possible to know simultaneously both the momentum, p, and the position, x, precisely.

A racquetball court is 40 feet long (about 12 m), and a ball is 2.25 inches in diameter and weighs 1.4 ounces (about 0.04 kg).

Clearly, racquetball is a game describable by classical mechanics. You can watch the ball bounce back and forth by observing it with light without changing it.

PARTICLE IN A BOX—QUANTUM

What are the differences if we now consider quantum racquetball? The court is still perfect, but now its length is 1 nm (10^{-9} m) rather than 12 m. Furthermore, the particle now has the mass of an electron, 9.1×10^{-31} kg rather than 0.04 kg. This is the quantum particle in a box problem.

We can immediately say that the lowest energy of a quantum particle in a 1-nm-length box cannot be zero. In the classical racquetball court, the velocity V could be zero, which means the momentum, p = mV is zero. In addition, the position x could have a perfectly well-defined value. For example, the ball could be standing still (V = 0) exactly in the middle of the court, which would be x = L/2. Then, for our classical racquet ball, $\Delta p = 0$ and $\Delta x = 0$. The product, $\Delta x \Delta p = 0$, is not in accord with the Heisenberg Uncertainty Principle, which is okay because this is a classical system. However, the absolutely small particle in the nanometer size box is a quantum particle, and it must obey the Uncertainty Principle, that is, $\Delta x \Delta p \geq h/4\pi$. If V = 0 and x = L/2, we know both x and p. The result would be $\Delta x \Delta p = 0$, the same as the classical racquetball. This is impossible for a quantum system. Therefore, V cannot be zero. The particle cannot be standing still at a specific point. If V cannot be zero, then E_k can never be zero. The Uncertainty Principle tells us that the lowest energy that a quantum racquetball can have cannot be zero. Our quantum racquetball can never stand still.

ENERGIES OF A QUANTUM
PARTICLE IN A BOX

What energies can a quantum particle in a nanometer-size box have? This question can be answered without a great deal of math, but we need to think about waves again. In Chapter 6, we discussed the wavefunctions for free particles. The wavefunction for a free particle with a definite momentum p is a wave that extends throughout all space. So an electron with a perfectly well-defined momentum is a delocalized wave over all space. The probability of finding a free electron is equal everywhere. Such an electron has a well-defined kinetic energy, $E_k = 1/2mV^2$, because it has a well-defined momentum, $p = mV$.

An electron in a nanometer-size box is something like our free particle in the sense that inside the box, $Q = 0$. Inside the box there is no potential, which means that there are no forces acting on the particle. This is just like the free particle; there are no forces acting on a free particle. However, there is a major difference between a particle in the box and a free particle, the walls of the box. An electron in a box is located only inside the box. Its wavefunction cannot be spread over all space because of the perfect nature of the box. The particle is inside the box and can't ever be outside the box. The wavefunction gives the probability amplitude of finding the particle in some region of space. This is the Born interpretation of the wavefunction. If our electron can only be found inside the box and never outside of the box, there must be finite probability of finding the particle inside the box but zero probability of finding the particle outside the box. If the probability of finding the particle outside the box is zero, then the wavefunction must be zero for all locations outside the box.

The result of the reasoning just presented is that the wavefunction for a particle in a box is like a free particle wavefunction, but the wavefunction must be zero outside the box. In his interpretation of the nature of the quantum mechanical wavefunction, Born placed certain physical constraints on the form wavefunctions can have. One of these is that a good wavefunction must be continuous. This condition means that the change in the wavefunction with position must be smooth. An infinitesimal change in position cannot produce a sudden jump in the probability. This is really a simple idea. If the probability of finding a particle in some very small region of space is, for example, 1%, then moving over an unimaginably small amount can't suddenly make the probability of finding the particle 50%. This is clear from the illustrations of wave packets in Figure 6.7. The probability changes smoothly with position. Therefore, we can say something else about the wavefunctions for a particle in the box in addition to the fact that they are waves with finite amplitudes inside the box and zero amplitude outside the box. Because the wavefunction must be continuous, right at the walls of the box the wavefunction inside the box must have zero amplitude so that it will match up with the wavefunction's zero amplitude outside of the box.

Figure 8.3 illustrates a discontinuous wavefunction (not allowed) inside a box. The wavefunction is called φ (Greek letter phi). The vertical axis gives the amplitude of the wavefunction. The dashed line shows where zero is. Wavefunctions, which are probability amplitude waves, can oscillate positive and negative. The wave-

FIGURE 8.3. *A wavefunction inside the box that is discontinuous. The wavefunction is called φ. The vertical axis is the amplitude of the wavefunction. The dashed line shows where the wavefunction is zero, which must be outside the box. The wavefunction has a nonzero value at the walls and then must drop discontinuously (not smoothly) to zero outside the box.*

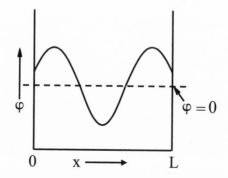

function shown in Figure 8.3 has values at the walls that are not 0. However, the wavefunction must be zero outside the box, that is, for values of x less than 0 and greater than L it must be zero. As drawn, the wavefunction jumps suddenly from nonzero values at the walls to zero values immediately beyond the walls outside the box. Therefore, the wavefunction as drawn in Figure 8.3 is not a good wavefunction because it is not continuous. This function cannot represent a quantum particle in the box.

Wave Function Must Be Zero at the Walls

For the wavefunctions representing the particle in the box to be physically acceptable functions, their values at the walls must be zero so that there is no discontinuity at the walls. This is not a difficult condition to meet. Figure 3.1 illustrates a wave in free space. It oscillates positive and negative. Every time it goes from positive to negative or negative to positive, it crosses through zero. In fact, the zero points are separated by one-half of a wavelength. So what we need to do to get good particle in a box wavefunctions is pick waves with wavelengths such that they fit in the box with their zero points right at the walls. Figure 8.4 shows three examples of waves that are acceptable particle in the box wavefunctions. The one on the bottom, labeled n = 1, is composed of a single half wavelength. It starts on the left with an amplitude of 0, goes through a maximum, and then is zero again at the wall at position L. The next wave up, labeled n = 2, is one full wavelength. Again, it starts at the left wall with amplitude zero, goes through a positive peak, back through zero, a negative peak, and is zero at the wall at position L. The wave labeled n = 3 is one and one-half wavelengths. Any wave that is an integer number of half wavelengths, that is 1, 2, 3, 4, 5, etc. half wavelengths, and has a wavelength so that it starts at zero on the right and ends at zero on the left is okay.

The label n is the number of half wavelengths in the particular

FIGURE 8.4. *Three examples of wavefunctions, φ, inside the box that are continuous. They have been shifted upward for clarity of presentation. The vertical axis is the amplitude of the wavefunction. The dashed line shows where the wave function is zero, which it must be outside the box. The wavefunctions, which have zero values at the walls, are continuous across the walls.*

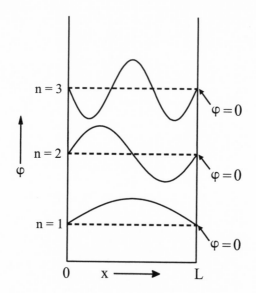

wavefunction. For n = 1, the wavelength, λ, is 2L because the box has length L, and n = 1 corresponds to a half wavelength. For n = 2, the wavelength is L because exactly one wavelength fits between the walls. For n = 3, 3 half wavelengths = L. That means 1.5λ = L. Then λ = L/1.5, so λ = 2L/3. Notice that there is a general rule here. λ = 2L/n, where n is an integer. For n = 1, λ = 2L. For n = 2, λ = 2L/2 = L. For n = 3, λ = 2L/3, and so forth.

Nodes Are Points Where the Wavefunction Crosses Zero

Nodes are another important feature of the wavefunctions. Nodes are points where the wavefunction crosses zero, going from positive to negative, or negative to positive. The n = 1 wavefunction has no nodes. The n = 2 wavefunction has one node right in the middle of the box. The n = 3 wavefunction has two nodes. Nodes are points (other than at the walls) where the probability of finding the particle is zero. In a classical system, as in Figure 8.2, the ball bounces back and forth. It can be at any location. But a particle in a quantum box has certain places (nodes) where the probability of finding it is zero. No matter how many measurements are made on identically prepared systems, we will never find the particle at a node.

Figure 8.4 shows the probability amplitude waves. As discussed, the probability of finding the particle in a certain region of space is proportional to the square of the wavefunction (actually the absolute value squared, but for our purposes, there is no difference). Figure 8.5 shows the square of the wavefunctions that are displayed in Figure 8.4. The square of the wavefunctions are always positive because the probability of finding a particle in some region of space cannot be negative. Where the amplitude is large, there is a large probability of finding the particle. As n increases, the number of nodes increases. As we will discuss in the next and later chapters, atomic and molecular wavefunctions also have nodes.

A question that is frequently asked is, how does a particle get through a node? For example, for n = 2, there is a node exactly in the middle of the box. In a classical system if we had a ball on the left side of the box and it was traveling to the right, but we said it could never be in the center of the box, we would be confident that the ball could never get to the right side of the box. However, we cannot think classically about an absolutely small particle, such as an electron in a molecular-sized box. It does not have a simultaneous definite position and momentum that can be described by an observable trajectory. A quantum particle, an electron, is described as a probability amplitude wave. Waves have nodes. Even classical waves have nodes. A quantum particle does not have to "pass

FIGURE 8.5. *The squares of the first three wavefunctions, φ², for the particle in a box. They have been shifted upward for clarity of presentation. The vertical axis is the amplitude of the wavefunction squared. The dashed line shows where the wavefunction is zero. The square of the wavefunctions are always positive because they represent probabilities. The wavefunctions shown in Figure 8.4 can be positive or negative.*

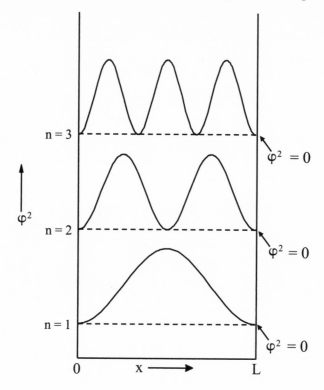

through" a node because it is a delocalized probability amplitude wave. The idea of a trajectory in which to get from point A to point B a particle must pass through all the points in between just doesn't apply to the proper wave description of electrons and other absolutely small particles.

The Energies Are Quantized

Now we will determine the possible energies that an absolutely small particle in a box can have. The classical ball in a racquetball court can have any energy, and the energy is continuous. We can determine what energies a particle, such as an electron, can have in a tiny box by using the rule for the possible wavelengths, $\lambda = 2L/n$, that allowed probability amplitude waves can have inside the box (see Figure 8.4). Here tiny means a box that is small in the absolute sense, that is, the wavelength is comparable to the size of the box. We will also need several other physical relationships that we have met previously. The other relations we need are the de Broglie wavelength, $p = h/\lambda$, where p is the momentum and h is Planck's constant; the fact that the momentum is $p = mV$, where m is the mass and V is the particle velocity; and the kinetic energy of the particle, $E = \frac{1}{2}mV^2$. Now let's combine these formulas.

First, square p. Then,

$$p^2 = m^2V^2.$$

If we now divide both sides of the equation by 2m, we see that the right side gives $\frac{1}{2}mV^2$ the kinetic energy, and the left side gives $\frac{p^2}{2m}$. So we have the following expression for the kinetic energy,

$$E = \frac{p^2}{2m}$$

Using the de Broglie relation, we can replace p^2 with $p^2 = h^2/\lambda^2$. Putting this into the expression for the energy gives,

$$E = \frac{h^2}{2m\lambda^2}.$$

Finally, we will use our rule, $\lambda = 2L/n$, for the possible wavelengths. Then $\lambda^2 = 4L^2/n^2$. Substituting this expression for λ^2 into the expression for the energy yields

$$E = \frac{n^2h^2}{8mL^2},$$

with n being any integer, 1, 2, 3, etc. The integer n is called a quantum number.

We have obtained a very important result, the energies for an absolutely small particle in an absolutely small box. The results are closely related to electrons in atoms or molecules. As can be seen in the formula, the energies are not continuous because n can only take on integer values; the other parameters are constants for a particular system. We say that the energy is quantized. It can only have certain values, which are determined by the physical properties of the system and the quantum number.

A Discreet Set of Energy Levels

There is a discreet set of energy levels for a given mass, m, and a given box length, L. As the quantum number n takes on values, 1, 2, 3, etc., the energies are

$$\frac{h^2}{8mL^2}, \frac{4h^2}{8mL^2}, \frac{9h^2}{8mL^2}, \text{ etc.}$$

Figure 8.6 is an energy level diagram for the first few energy levels of the particle in a box. The energy is plotted in units of $h^2/8mL^2$. To get an actual energy, it is only necessary to plug in particular values for m and L in the energy level formula. The plot shows the energy increasing as the square of the quantum number n. The dashed line locates where the energy is zero. In the quantum particle in a box, the lowest energy level does not have zero energy, in contrast to a classical particle in a box. In the classical racquetball court, the energy that the ball can have is continuous. By hitting the ball a little harder or slightly softer, the ball's energy can be changed any amount up or down. Here, the quantum racquetball can only

FIGURE 8.6. *Particle in a box energy levels. The quantum number is n. E is the energy, which increases as the square of the quantum number. The energy is plotted in units of $h^2/8mL^2$, so that it is easy to see how the energy increases. The dashed line is zero energy. The lowest energy level does not have E = 0, in contrast to a classical particle in a box.*

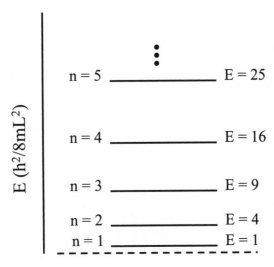

take on energies that have distinct values, as shown in Figure 8.6. As we discussed at the beginning of our analysis of the quantum particle in a box, the lowest energy is not zero. If the quantum particle in a box could have zero energy, it would violate the Uncertainty Principle.

PARTICLE IN A BOX RESULT
RELATED TO REAL SYSTEMS

The particle in a box is a very simple example of a general feature of absolutely small systems. The energy of such systems is not necessarily continuous. The particle in the box is not a physically realizable system because it is one dimensional and it has "perfect" walls.

However, atoms and molecules are real systems. The energy levels of atoms and molecules have been studied in great detail, and their quantized energy levels have been measured and calculated. Just as the energy levels of the particle in the box depend on the properties of the system, that is, the mass of the particle and the length of the box, the energy levels of atoms and molecules depend on the properties of the atoms and molecules.

Molecules Absorb Light of Certain Colors

Although the particle in a box is not a physically realizable system, features of this problem are also found in atoms and molecules. In the photoelectric effect, the incident photon energy is so great that electrons fly out of the piece of metal (see Chapter 4). For high enough energy, a photon incident on a molecule can also result in electron emission. However, for lower energy photons, when light shines on an atom or molecule, it can be absorbed without electron emission. The atom or molecule will have its internal energy increased because it has the additional energy of the photon. Molecules (and atoms) are composed of charged particles, electrons that are negatively charged, and atomic nuclei that are positively charged. In the visible and ultraviolet range of wavelengths of light, that is, wavelengths shorter than 700 nm, the frequency of light is very high. The oscillating electric field of the light interacts with the charged particles of the molecules. Electrons are very light, and therefore, it is easy for them to respond to the rapidly oscillating electric field of visible or ultraviolet light. The absorption of visible or ultraviolet light is caused by increasing the energy of the electrons in a molecule.

The question is, what wavelengths of light will be absorbed by a molecule? This is a very complex question for any given molecule. Large quantum theoretical calculations are performed to determine the absorption spectrum of a molecule. However, we can learn about important aspects of molecular absorption of light from the particle in a box problem. As an exceedingly simple model of a molecule, we will consider a single electron in a molecular-sized box. Later we will put in numbers. The electron will be in its lowest energy state, called the ground state, when no light shines on the electron in the box (the molecule). For the particle in the box, the lowest energy has the quantum number, n = 1. For n = 1, the energy is:

$$E = \frac{h^2}{8mL^2}.$$

When light shines on a molecule, a photon can be absorbed. If the photon is absorbed, the energy of the photon is lost from the total energy of all of the light. Energy must be conserved, which happens by an electron going into a higher energy state, that is, it goes from the ground state, the lowest energy level, to a higher energy level. However, this higher energy level cannot have any energy value because the energy levels of the particle in a box (and molecules) are quantized. The lowest energy state above the ground state has quantum number n = 2. This is called an excited state. The electron has been excited by absorption of a photon from the ground state to the first excited state. The energy of the first excited state, the n = 2 state, is:

$$E = \frac{4h^2}{8mL^2}.$$

Energy must be conserved. This is true in classical mechanics, and it is true in quantum mechanics. We start with the electron in the ground state. When the photon is absorbed, the electron is in an excited state. Therefore, to conserve energy, the photon energy must equal the difference between the excited state energy and the ground state energy. Only a photon with this energy can be absorbed by the system. The photon energy determines the wavelength of the light. Therefore, only certain colors of light can be absorbed.

Figure 8.7 illustrates absorption of a photon. The arrows show two allowed paths for photon absorption. These are called transitions. The transitions from n = 1 to n = 2, and n = 1 to n = 3 are shown in the figure. For a photon to be absorbed, the photon energy must equal the difference in energy of two of the quantum levels. If the photon energy does not match the difference in energy between two levels, it cannot be absorbed.

The difference in the energy, ΔE, between the n = 2 first excited state energy level and the n = 1 ground state energy level is:

FIGURE 8.7. *Particle in a box energy levels. The quantum number is n. E is the energy plotted in units of $h^2/8mL^2$. The arrows indicate absorption of photons that can take an electron from the lowest energy level, n = 1, to higher energy levels, n = 2, n = 3, etc. For a photon to be absorbed, its energy must match the difference in energy between two energy levels.*

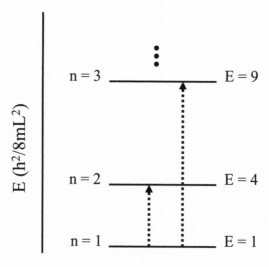

$$\Delta E = \frac{4h^2}{8mL^2} - \frac{h^2}{8mL^2}$$

$$\Delta E = \frac{3h^2}{8mL^2}.$$

This is the energy that the photon must have to cause the electron to make a transition from the ground state to the first excited state. We can use Planck's relation, $E = h\nu$, for the photon energy to see that the energy ΔE corresponds to a certain frequency of light. Also, because $\lambda\nu = c$, the wavelength times the frequency equals the speed of light, we can determine the wavelength (color) of the light that will be absorbed.

The Color of Fruit

Let's put in numbers. $h = 6.6 \times 10^{-34}$ J-s. The electron mass, $m_e = 9.1 \times 10^{-31}$ kg. For the length of the box, let's take L to be that of a medium-sized molecule, that is, $L = 0.8 \times 10^{-9}$ m (0.8 nanometers, 0.8 nm). Then,

$$\Delta E = \frac{3(6.6 \times 10^{-34})^2}{8(9.1 \times 10^{-31})(0.8 \times 10^{-9})^2} = 2.8 \times 10^{-19} \text{J}.$$

Converting this energy to a frequency by dividing by h gives $\nu = 4.25 \times 10^{14}$ Hz, which corresponds to the wavelength of the light that will be absorbed, $\lambda = 7.06 \times 10^{-7}$ m = 706 nm. Light with wavelength 706 nm is very deep red. It is on the red edge of colors that you can see by eye. What happens if the box (molecule) is smaller, say 0.7 nm instead of 0.8 nm? The energy of the light that is absorbed becomes higher and, therefore, the wavelength of light that is absorbed becomes shorter as the box becomes smaller. The energy absorbed goes inversely as L^2 (L^2 is in the denominator), which means as the box gets smaller, the energy levels get further apart, and the energy difference increases as the square of the box length. So for a 0.7 nm box, the absorbed wavelength is $\lambda = 540$ nm, which is green light. If the box is even smaller, say 0.6 nm, then $\lambda = 397$ nm, which is very blue, on the blue edge of the light that can be seen with the unaided eye.

These simple results are also generally true for molecules, although there are a lot of details that come into play. However, for a sequence of molecules that have essentially the same type of structure (types of atoms, etc.) the bigger the molecule, the further to the red it will absorb. The results presented for a particle in a box show very qualitatively why things have different colors. Small molecules absorb light in the ultraviolet part of the spectrum. We can't see ultraviolet light, so absorption by small molecules does not result in a color. We see colors by the light that bounces off an object. The colors that are absorbed don't bounce off. Large molecules absorb in the visible part of the spectrum, and these molecular absorptions give objects color.

Cherries are red and blueberries are blue because they have different molecules in them that absorb different colors of light strongly. These molecules have quantized electronic transitions. They can only absorb light from their ground electronic states to their excited states at wavelengths that are determined by their quantized energy levels. In the particle in a box, the transition energies of an electron are determined solely by the length of the box and the mass of the electron. For molecules, the quantized transition energies, and therefore wavelengths and colors, are determined both by the sizes of the molecules and by the details of the molecular structures, that is, the shape of the molecule, the types of atoms that make up the molecule, and how the atoms are arranged. Dyes are molecules that have specific strong absorptions in the visible. Dyes are used to give our clothes different colors. Brightly colored plants, green leaves and red roses, contain a wide variety of molecules with different sizes and shapes that absorb distinct colors of light strongly. The sizes and shapes of these molecular absorbers give the plants their brilliant colors. If the molecules absorb green and red strongly, then blue will bounce off of an object, and it will look blue. If blue and green are absorbed strongly, mainly red will bounce off, and an object will look red. The colors that are absorbed are determined by the quantized energy levels of the molecules in the object.

We observe color constantly in everyday life. Color is one of many phenomena that we encounter that is inherently quantum mechanical. There are many others. For example, when you turn on an electric stove, the element gets hot. Why electricity moving through metal produces heat (Chapter 19) is another everyday quantum phenomenon. Why is carbon dioxide a greenhouse gas (Chapter 17)? What is a trans fat (Chapter 16)? It is necessary to go into the quantum mechanics of molecular structure to understand such systems. In the following chapters, the quantum description of atoms and molecules will be developed and applied to a number of common issues and problems. The necessary machinery for understanding atoms and molecules is developed in Chapters 9 through 14. These chapters supply a great deal of interesting information about how atoms and molecules behave and provide the bridge between the general ideas of quantum theory that we have just developed and understanding many phenomena that surround us.

9

The Hydrogen Atom:
The History

IN CHAPTER 8, WE DISCUSSED the particle in a box problem. We imagined an electron confined to a very small one-dimensional box, as shown in Figure 8.1. The particle in a box is a useful problem because the math is simple enough to find the quantized energy levels without great difficulty. A formula was obtained that showed that the energy states of the particle in a box come in discreet steps that depend on a quantum number n, where n is an integer that starts at 1 and can take on any integer value. However, it was pointed out that this is a very artificial example of quantum confinement. In nature, there are no truly one-dimensional systems. Furthermore, the walls of the box are infinitely high and completely impenetrable. This is also physically unrealistic. As discussed in connection with the photoelectric effect in Chapter 4, if a photon has sufficient energy to overcome the binding energy of electrons to the atoms in a piece of metal, the interaction of the photon with an initially bound electron can eject the electron from the metal (see Figure 4.3).

However, for a number of reasons it is very useful to examine the particle in a box. First, we found that the energy levels are quantized (see Figure 8.6). In contrast to classical mechanics, the energy an electron can have when confined to a box the size of an atom or molecule is not continuous. Rather, the energy comes in discrete steps. A photon of the right energy can excite an electron from one energy level to another (see Figure 8.7). The energy of the photon must match the difference between the energy of the level the electron starts in and the energy of the one it ends up in. But in contrast to real systems, no amount of energy can cause the electron to come flying out of the box because the walls are infinitely high. This is a way of saying that an electron would have to have infinite energy to get out of the box. The box is an infinitely deep well and the electron is trapped in it; no finite amount of energy can overcome the infinite binding energy.

Another important feature of the particle in the box is the nature of the wave functions. The wave functions are probability amplitude waves that are related to where the electron is inside of the box (see Figure 8.4). The square of these wavefunctions (Figure 8.5) gives the probability of finding the electron in some region of space. The probability amplitude waves have nodes. As the quantum number increases, the number of nodes increases, as can be seen in Figure 8.5. Nodes are places where the probability of finding a particle, such as an electron, is zero.

Atoms are real three-dimensional physical systems in contrast to the one-dimensional particle in the box. The three-dimensional nature of atoms is a major difference, but as discussed in Chapter 10, some of the most important features of the quantum mechanical description of atoms are qualitatively similar to the particle in the box results. Atoms have quantized energy levels. They have wavefunctions that have an increasing number of nodes as the quantum number increases. Many other things are very different. The quantum states of atoms have associated with them several quantum numbers, and because atoms are three dimensional, their wavefunctions have three-dimensional shapes. These properties of atoms will be discussed in detail beginning in Chapter 10 with the simplest atom, the hydrogen atom. But first, we will look at some of the early observations that indicated that classical mechanics was not going to be able to describe atoms.

THE SOLAR BLACK BODY RADIATION SPECTRUM

We have introduced the experimental method of spectroscopy, that is, taking a spectrum of the light that comes out of a system or the spectrum of the light that is absorbed by a system. A spectrum is just a recording of the intensity of the various colors of light. We measure the amount of light at each wavelength (color). When we refer to colors, we don't mean only the colors we can see, the visible spectrum, but also longer wavelengths (lower energy), the infrared, and shorter wavelengths (higher energy), the ultraviolet. A system may be a container of gas molecules, the leaf of a plant, or molecules in a liquid such as the molecules that make wine red. We use complicated dye molecules to color our clothes because the size and structure of a molecule determines which wavelengths of light it will absorb.

In Chapter 4, black body radiation was briefly discussed. A hot object gives off light. A fairly hot piece of metal will glow red. This can be seen in the wire elements of an electric space heater or an electric stove. As the temperature is increased the color will move toward the blue. It was mentioned that stars are well described as black bodies, and the color of a star can be used to determine its temperature. Planck developed a formula that yields the black body spectrum for a given temperature. Figure 9.1 shows the solar spectrum calculated using Planck's formula that is the closest to the experimentally measured solar spectrum. The frequency is plotted in wave numbers (cm^{-1}). Multiplying the frequency in cm^{-1} by the speed of light in centimeters per second (3×10^{10} cm/s) gives the

FIGURE 9.1. *The black body spectrum of the sun calculated using the Planck formula for black body radiation from a hot object. The curve is a good representation of the solar spectrum without some of the fine details. The lower axis is the frequency in wave numbers (see text). The top axis is the wave length in nanometers. The green light is 500 nm. Very blue is 400 nm; very red is 666 nm. The vertical axis is the amount of light (see text).*

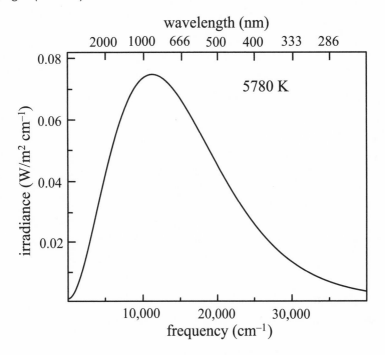

frequency in Hz, the conventional frequency units. The top axis in Figure 9.1 is the wavelength in nanometers (nm). 500 nm is green light. 400 nm is very blue light. 666 nm is very red light. 333 nm is in the ultraviolet region of the optical spectrum and cannot be seen with the eye. 1000 nm is in the infrared part of the spectrum and also cannot be seen with the eye. These wavelengths can be detected with electronic photodetectors. Originally, they were detected with photographic film. The vertical axis is the irradiance. It is the number of watts (joules per second) that would fall on an area of a square meter in a little slice of frequency 1 cm^{-1} wide. Basically the plot is the amount of energy per second of a particular color that falls on a square meter.

The shape of the spectrum plotted in Figure 9.1 is almost identical to the actual solar spectrum. The calculated spectrum is obtained by adjusting the temperature in Planck's formula until the best match is obtained to the experimental spectrum. The temperature that gives the best spectrum is 5780 K, where K is the unit of temperature in degrees Kelvin. This is the unit of absolute temperature developed by William Thomson, the first Baron Kelvin (Lord Kelvin, 1824–1907). The Kelvin scale is used in physics and chemistry because it has a well-developed physical meaning for 0 K, the absolute zero of temperature. At 0 K, all atomic motions associated with kinetic energy (heat), the energy of moving particles, stop. To obtain the temperature in degrees centigrade (C) subtract 273. Then in centigrade, the sun's temperature is 5507 C. To convert to Fahrenheit (F), multiply by 9/5 and add 32 to the temperature in centigrade. Therefore, the surface temperature of the sun is 9945 F. The Fahrenheit temperature scale is named for Daniel Gabriel Fahrenheit (1686–1736).

Dark Lines in the Solar Spectrum

It is remarkable that the Planck formula developed using the first quantum concept, that the energies of electrons "oscillating" in a metal are not continuous, can be used to determine the temperature of stars. The calculated spectrum shown in Figure 9.1 is continuous because a hot object produces a continuous distribution of colors (energies of light). While the experimental measurement has the shape shown in Figure 9.1, it also has some very sharp features in it that are not part of the sun's black body spectrum. Figure 9.2 is

FIGURE 9.2. *The visible portion of the solar spectrum. The continuous range of colors is the black body spectrum. The dark lines or bands are colors that do not reach Earth, so they appear as colors missing from the solar spectrum. The wavelengths of the lines and the spectrum are given in nm, nanometers, which are billionths (10⁻⁹) meters.*

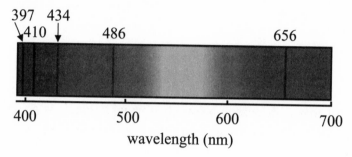

an illustration of the solar spectrum with thin dark lines that represent a lack of light at certain frequencies. The spectrum shown in Figure 9.1 is the light that comes from the sun. But the dark lines are certain narrow bands of color that do not make it to earth. The light is being absorbed between the sun and the earth. The lines are called absorption lines or bands. These same lines are very prominent in the spectra of light coming from stars other than our own sun.

The same wavelengths that are seen as dark lines in the solar spectrum can also be seen as distinct colors from an arc lamp filled with hydrogen gas. A hydrogen arc lamp or discharge lamp is a sealed glass cylinder filled with hydrogen gas with electrodes at either end. When a sufficiently high voltage is connected to the lamp, positive connected to one electrode and negative to the other electrode, electricity arcs through the lamp like a small continuous lightning bolt. The colors (wavelengths) in the visible coming out of the lamp are the same as wavelengths of the black lines shown in Figure 9.2.

The Hydrogen Line Spectrum

The first attempt at understanding the line spectrum of hydrogen in the visible region was made in 1885 by the Swiss schoolteacher and mathematician, Johann Balmer (1825–1898). Balmer noted that the frequencies of the lines, f, in the visible part of the spectrum were related by the formula

$$f \propto \frac{1}{2^2} - \frac{1}{n^2}.$$

The symbol $\propto$ means proportional to, so there is a multiplicative constant that is discussed below. In this equation, n is an integer greater than 2, that is, 3, 4, 5, etc. The spectral lines in the visible are called the Balmer series.

Later, lines were discovered in the ultraviolet and the infrared. These are called the Lyman series and the Paschen series, respectively, after their discoverers Theodore Lyman (1874–1954), a U.S. physicist and spectroscopist, and Louis Karl Heinrich Friedrich Paschen (1865–1947), a German physicist. In 1888, the Swedish physicist and spectroscopist, Johannes Rydberg (1854–1919) presented a formula that described all of the spectral lines seen in emission from a hydrogen arc lamp or in the absorption spectrum of solar or stellar light. The Rydberg formula for the frequency of the hydrogen atom spectral lines is

$$f = R_H \left(\frac{1}{n_1^2} - \frac{1}{n_2^2} \right).$$

n_1 is an integer beginning at 1. n_2 is another integer that must be greater than n_1. $n_1 = 1$ gives the Lyman series. $n_1 = 2$ gives the Balmer series. $n_1 = 3$ gives the Paschen series. The constant, R_H, is called the Rydberg constant for the hydrogen atom. It has the value, $R_H = 109,677$ cm^{-1}. Here the constant is given in wave numbers (cm^{-1}). When this value is used in the Rydberg formula, the frequency of a spectral line determined by the integers n_1 and n_2 is in wave numbers. To get it in Hz, the result is multiplied by the speed of light in cm/s, that is, 3×10^{10} cm/s. To find the wavelength of a spectral line, take the inverse of the frequency in wave numbers, that is, take 1 and divide it by the frequency in wave numbers. For example, if $n_1 = 2$ and $n_2 = 3$, then

$$f = R_H \left(\frac{1}{2^2} - \frac{1}{3^2} \right) = R_H \left(\frac{1}{4} - \frac{1}{9} \right) = 1.52 \times 10^4 \text{ cm}^{-1},$$

the frequency in wave numbers. The inverse of this number is 6.56×10^{-5} cm $= 656 \times 10^{-9}$ m. 10^{-9} m is a nanometer, so the wavelength is 656 nm. This is the red line in the Balmer series shown in Figure 9.2.

In connection with Figure 8.7, we already discussed discreet optical transitions between quantized energy levels for the particle in a box. Figure 8.7 shows transitions between the particle in a box states for n = 1 going to n = 2 and n = 1 going to n = 3. So it should come as no surprise that the optical transitions of the hydrogen atom could involve discreet frequencies that depend on integers. However, in 1888, at the time of the Rydberg formula, it was still 12 years before the first use of the idea of quantized energy levels by Planck to explain black body radiation, and 37 years before true quantum theory took shape in 1925. The various series of spectroscopic lines that have energies related to integers through the Rydberg formula can be understood as optical transitions between discreet energy levels that are associated with the hydrogen atom. A few of the energy levels that give rise to the Lyman series and the Balmer series are shown in Figure 9.3. In the figure, the downward arrows indicate the emission of light that would come from a hydrogen arc lamp. The hydrogen atom starts in a higher energy level and ends up in a lower energy level. Energy is conserved by the emission of a photon. To conserve energy, the photon must have the energy difference between the initial higher energy level and the final lower energy level. In the Rydberg formula, the smallest value

FIGURE 9.3. *Schematic of some of the energy levels that give rise to the Lyman and Balmer series of hydrogen atom emission lines. The down arrows indicate that light is being emitted from a hydrogen arc lamp, for example. For absorption, shown by the black lines in Figure 9.2, the arrows would point up. The level spacings are indicative but not to the true scale.*

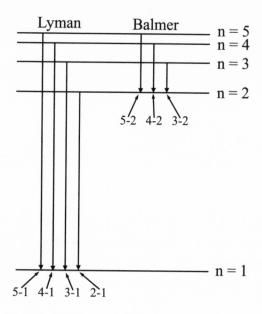

n_1 can have is 1, and n_2 must be bigger than n_1. The arrow labeled 2-1 represents emission from the n = 2 level to the n = 1 level. The next higher energy emission in the Lyman series is for emission from the n = 3 level to the n = 1 level. In the Rydberg formula, the next possible value for n_1 is 2, and n_2 must be bigger than n_1. Therefore, the lowest energy emission line in the Balmer series is labeled 3-2. The hydrogen atom begins in the n = 3 level and ends in the n = 2 level, and energy is conserved by emission of a photon with wavelength 656 nm. When light shines on hydrogen atoms, absorption occurs, which would be indicated in the diagram by up arrows. The Balmer series absorptions are shown in Figure 9.2.

BOHR'S HYDROGEN ATOM THEORY— NOT QUITE THERE

The first detailed description of the energy levels of the hydrogen atom was developed by Niels Bohr (1885–1962) in 1913. Bohr won the Nobel Prize in Physics in 1922 "for his services in the investigation of the structure of atoms and of the radiation emanating from them." Bohr's theory of the hydrogen atom is referred to as the old quantum theory. Bohr made many advances and in fact was able to calculate precisely the energy levels of the hydrogen atom, and therefore obtain the Rydberg relation and predict all of the hydrogen atom spectral lines. Bohr was also the first to propose two ideas we have already been using. He said an atomic system can only exist in certain states, which he called "stationary" states. Now, we usually refer to these as energy eigenstates. Each of these states has a well-defined energy, E. Transitions from one stationary state to another can occur by absorption or emission of light or other means that can give or take energy from the system, and the amount of energy must be equal to the difference in energy of the two states. This idea is the basis for Figures 9.3 and 8.7, where the arrows represent transitions between states that occur by absorption or emission of light.

Bohr also put forward what came to be known as the Bohn Frequency Rule. The frequency of light emitted or absorbed in making a transition from an initial state with energy E_1 to a final state with energy E_2 is the difference in the energies divided by Planck's constant, that is,

$$\nu = \frac{|E_1 - E_2|}{h}.$$

ν is the frequency and h is Planck's constant (h $= 6.6 \times 10^{-34}$ J-s). The vertical lines are the absolute value. For absorption E_1 is less than E_2, so $E_1 - E_2$ would be negative. The absolute value means that you make the number positive even if the difference is negative. The frequency, ν, has to be a positive number. Multiplying both sides by h gives E, the energy difference between the energy levels (stationary states) as E $= h\nu$, which is the Planck relationship that Einstein used to explain the photoelectric effect discussed in Chapter 4.

What is a hydrogen atom and what are the failures of the Bohr method? A hydrogen atom is composed of two charged particles, a proton that has a positive charge, that is, a charge of $+1$, and an electron that has a negative charge, a charge of -1. When we say a charge of 1, it actually is shorthand for the charge that is on one proton. The charge in real units is 1.6×10^{-19} C, where C is the Coulomb, the unit of charge. Ernest Rutherford (1871–1937) did experiments in 1911 that showed that atoms were composed of a small, heavy positively charged nucleus and one or more electrons outside of the nucleus. Rutherford won the 1908 Nobel Prize in Chemistry "for his investigations into the disintegration of the elements, and the chemistry of radioactive substances." Rutherford's findings applied to the hydrogen atom mean that the proton is the nucleus and a single electron is found outside the nucleus. Even for hydrogen where the nucleus is composed of a single proton, the nucleus is much heavier than an electron. The mass of a proton is $m_p = 1.67 \times 10^{-27}$ kg, while the mass of the electron is only $m_e = 9.1 \times 10^{-31}$ kg. A proton weighs about 1836 times as much as an electron.

In the Bohr model, the electron orbited the proton like a planet orbiting the sun. The lowest energy state of the hydrogen atom, n $= 1$, has the electron going around the proton in a circle. Higher energy states of the electron, with n greater than 1, could have different shapes. Some were still circles, but some were ellipses. This picture of the electron orbiting the proton should immediately set off danger warnings based on the material covered in earlier chapters. In Chapter 6, the Heisenberg Uncertainty Principle was discussed. We know that an absolutely small particle cannot have a classical trajectory. To have a trajectory, it is necessary to know the position and the momentum of a particle simultaneously over all time. But the Heisenberg Uncertainty Principle says that it is not possible to know both the position and the momentum precisely simultaneously. The uncertainty relation states that $\Delta x \Delta p \geq h/4\pi$, where h is Planck's constant. Absolutely small particles are described in terms of probability amplitude waves, not trajectories. Of course in 1913, when Bohr came out with his mathematical treatment of the hydrogen atom, the nature of absolutely small particles was not known.

The failure of Bohr's approach became apparent when it was applied to systems other than the hydrogen atom. While Bohr's method could predict very accurately the energy levels, and therefore the spectrum of the hydrogen atom, it could not do so for the next simplest atom, the helium atom. Nor could it properly predict the properties of the simplest molecule, the hydrogen molecule, which is composed of two hydrogen atoms. The Bohr method could not account for the strength of the chemical bond that held the two hydrogen atoms together to form the hydrogen molecule. Although Bohr made giant steps in the right direction, the failures of his approach ultimately led to the development of true quantum theory in 1925.

10

The Hydrogen Atom: Quantum Theory

In 1925 Schrödinger and Heisenberg separately developed quantum theory. Their two formulations are mathematically different, but their theories are rigorous and form the underpinning of modern quantum theory. At about the same time, Dirac made major contributions as well. First, he presented a unified view of quantum theory that showed that the Schrödinger and Heisenberg theories, while mathematically different, were equivalent representations of quantum mechanics. In addition, he developed a quantum theory for the hydrogen atom that is also consistent with Einstein's Theory of Relativity. The formulation by Schrödinger is the most often used to describe atoms and molecules. Therefore, most of our discussions, starting with the hydrogen atom and then going on to larger atoms and molecules, will be based on the concepts and language that is inherent in the Schrödinger approach.

THE SCHRÖDINGER EQUATION

We used a very simple but correct mathematical method for obtaining the energy levels of the particle in the box and the wavefunctions, but the method we used is not general. For example, it cannot be used to find the energy levels and the wavefunctions for the hydrogen atom. In fact, the language we have been using, that is, wavefunctions and probability amplitude waves, comes from Schrödinger's formulation of quantum theory. In 1925 Schrödinger presented what has come to be known as the Schrödinger Equation. The Schrödinger Equation is a complicated differential equation in three dimensions. We will not do the mathematics necessary to solve the Schrödinger equation for the hydrogen atom or other atoms or molecules. However, we will use many of the results to learn about atoms and molecules, beginning with the hydrogen atom.

The solution of the hydrogen atom problem using the Schrödinger Equation is particularly important because it can be solved exactly. The hydrogen atom is a "two-body" problem. There are only two interacting particles, the proton and the electron. The next simplest atom is the helium atom, which has a nucleus with a charge of +2 and two negatively charged electrons. This is a three-body problem that cannot be solved exactly. In classical mechanics, it is also not possible to solve a three-body problem. The problem of determining the orbits of the Earth orbiting the Sun with the Moon orbiting the Earth cannot be solved exactly with classical mechanics. However, in both quantum mechanics and classical mechanics, there are very sophisticated approximate methods that permit very accurate solutions to problems that cannot be solved exactly. The fact that a method is approximate does not mean it is inaccurate. Nonetheless, because the hydrogen atom can be solved exactly with quantum theory, it provides an important starting point for understanding more complicated atoms and molecules.

WHAT THE SCHRÖDINGER EQUATION TELLS US ABOUT HYDROGEN

What does the solution to the Schrödinger Equation for the hydrogen atom give? It gives the energy levels of the hydrogen atom, and it gives the wavefunctions associated with each state of the hydrogen atom. The wavefunctions are the three-dimensional probability amplitude waves that describe the regions of space where the electron is likely to be found. Schrödinger's solution to the hydrogen atom problem gives energy levels consistent with the empirically obtained Rydberg formula. The energy levels are

$$E_n = -\frac{R_H}{n^2},$$

where n is the principal quantum number. It is an integer that can take on values ≥1, that is, greater than or equal to 1. The difference in energy between any two energy levels is the Rydberg formula. However, in the Schrödinger solution, R_H is not an empirical parameter. In solving the problem, Schrödinger found the Rydberg constant is determined by fundamental constants, with

$$R_H = -\frac{\mu e^4}{8\epsilon_o^2 h^2}.$$

h is Planck's constant. e is the charge on the electron. ϵ_o is a constant called the permittivity of vacuum. $\epsilon_o = 8.54 \times 10^{-12}$ C²/J m, with the units Coulombs squared per Joule – meter. μ is the reduced mass of the proton and the electron. It is

$$\mu = \frac{m_p m_e}{m_p + m_e},$$

where m_p and m_e are the masses of the proton and electron, respectively. The charge on the electron and the proton and their masses were given above.

While Rydberg took experimental data and developed an empirical formula that described the line spectra of the hydrogen atom, the results of Schrödinger's solution to the hydrogen atom problem using quantum theory are fundamentally different. We have to spend a moment to marvel at the triumph of quantum theory that emerged in 1925. There are no adjustable parameters in Schrödinger's derivation of the energy levels of the hydrogen atom. All of the necessary constants are fundamental properties of the particles and the electrostatic interaction that attracts the electron's negative charge to the proton's positive charge. Schrödinger did not look at the experimental data and then adjust a constant, R_H, until it fit the data. He set up a theoretical formalism and applied it to the hydrogen atom. The application of his theory accurately reproduced the experimental observables, the hydrogen atom line spectra, using only fundamental constants. In contrast to Bohr's theory, the Schrödinger Equation has been successfully applied to a tremendous number of other problems including atoms other than hydrogen and small and large molecules. As mentioned above, for systems larger than the hydrogen atom, that is, atoms and molecules involving more than two particles, the Schrödinger Equation cannot be solved exactly. However, many powerful approximation techniques have been developed that enable accurate solutions to the Schrödinger Equation for atoms, molecules, and other types of quantum mechanical problems. With the advent of computers, and the current tremendous power of computers, it is possible to solve the Schrödinger Equation for very large and complex molecules. As discussed in subsequent chapters, molecules have shapes. The solutions to the Schrödinger Equation for a molecule give its energy levels and its wavefunctions. The wavefunctions provide the necessary information to describe the shapes of molecules.

THERE ARE FOUR QUANTUM NUMBERS

The energies of the different states of the hydrogen atom only depend on a single quantum number, n. However, there are actually four quantum numbers associated with electrons in atoms. These come out of solving the hydrogen atom with quantum theory. One of these only comes into play for atoms and molecules that have more than one electron. In that sense, the hydrogen atom is a special case because it has only one electron. In the hydrogen atom, in addition to the principal quantum number n, the two other quantum numbers are l and m. l is called the orbital angular momentum quantum number and m is called the magnetic quantum number. These two quantum numbers, when combined with n, determine how many different states are associated with a particular energy, and they determine the shapes of the wavefunctions. The fourth quantum number is s. It is called the spin quantum number. When Bohr solved the hydrogen atom problem with old quantum theory, the electron moved in orbits that had different energies and shapes. Schrödinger's correct quantum solution to the hydrogen atom gave the energies and the wavefunctions, which, in correspondence to Bohr's orbits, are called "orbitals." In discussing atoms and molecules, we often use the term wavefunction and orbital interchangeably. The orbitals are probability amplitude waves that obey Heisenberg's Uncertainty Principle, in contrast to Bohr's orbits.

As stated above, the principal quantum number, n, can take on values, $n \geq 1$, that is, 1, 2, 3, 4, etc. l can have values from 0 to $n - 1$ in integer steps. m can take on values from l to $-l$ in integer steps. s can only take on two values, $+1/2$ or $-1/2$. These values are summarized in Table 10.1.

For historical reasons, the states with different quantum numbers l are given different names. An s orbital has $l = 0$. A p orbital has $l = 1$. A d orbital has $l = 2$. An f orbital has $l = 3$. For our discussions of all atoms, we will only need to go to f orbitals, that is, $l = 3$. As shown below, these different orbital types have different shapes.

Because the energies of the states (orbitals) of the hydrogen atom only depend on the quantum number n, there will be more than one state with the same energy for $n > 1$. For $n = 1$, $l = 0$,

TABLE **10.1.** *Quantum Numbers.*

$$n \Longrightarrow 1, 2, 3, 4, 5, \cdots$$

$$l \Longrightarrow 0 \text{ to } n-1 \text{ (integer steps)}$$

$$m \Longrightarrow +l \text{ to } -l \text{ (integer steps)}$$

$$s \Longrightarrow +1/2 \text{ or } -1/2$$

$$\left[\begin{array}{ll} l = 0 & s \\ l = 1 & p \\ l = 2 & d \\ l = 3 & f \end{array}\right.$$ called orbital or state

s only if more than one electron

and m = 0 (see the table). Therefore, there is a single orbital with n = 1. It has $l = 0$, so it is referred to as the 1s orbital. The 1 is the n value, and s means that $l = 0$. For n = 2, l can equal 0, giving rise to the 2s orbital. However, for n = 2, l can also equal 1. For l = 1, m can equal 1, 0, or −1 (see the table). $l = 1$ is a p orbital, and there are three different p orbitals that can be called $2p_1$, $2p_0$, and $2p_{-1}$. The 2 is the principal quantum number n. The p means $l = 1$, and the three subscripts are the three possible m values. So for n = 2, there are four different states.

If n = 3, then l can equal 0, to give the 3s orbital. l can also equal 1 with m = 1, 0, and −1, to give $3p_1$, $3p_0$, and $3p_{-1}$. In addition, l can equal 2. For $l = 2$, m can equal 2, 1, 0, −1, and −2. These are the d orbitals, $3d_2$, $3d_1$, $3d_0$, $3d_{-1}$, and $3d_{-2}$. There are five different d orbitals. Therefore, for n = 3, there are nine different states: an s orbital, three p orbitals, and five d orbitals. When n = 4, there is the 4s orbital, the three different 4p orbitals, $4p_1$, $4p_0$, and $4p_{-1}$, the five different 4d orbitals, $4d_2$, $4d_1$, $4d_0$, $4d_{-1}$, and $4d_{-2}$. In addition, there are seven f orbitals, $4f_3$, $4f_2$, $4f_1$, $4f_0$, $4f_{-1}$, $4f_{-2}$, and $4f_{-3}$. Therefore, for n = 4, there are a total of 16 states: an s orbital, three p orbitals, five d orbitals, and seven f orbitals.

As mentioned above, each of the orbitals has a different shape. It is common to rename the orbital with an indication of its shape. For example, the three different 2p orbitals, rather than being called

$2p_1$, $2p_0$, and $2p_{-1}$, are usually called $2p_x$, $2p_z$, and $2p_y$. The relation between the subscript and the shape will become clear when the shapes are presented.

HYDROGEN ATOM ENERGY LEVELS

Figure 10.1 shows an energy level diagram for the hydrogen atom. Levels are shown for n = 1 to 5. The spacings between the levels are not scaled properly for clarity of presentation, but as shown, the spacing between levels gets smaller as n increases. Also as n in-

FIGURE 10.1. *Hydrogen energy level diagram. The spacings between the levels are not to scale. The first five energy levels are shown. The energy only depends on the principal quantum number, n. The orbitals and the number of each type are also shown. For n = 4, there is a single s orbital, three different p orbitals, five different d orbitals, and seven different f orbitals. The diagram would continue with the n = 6 level. The different levels are sometimes referred to as shells.*

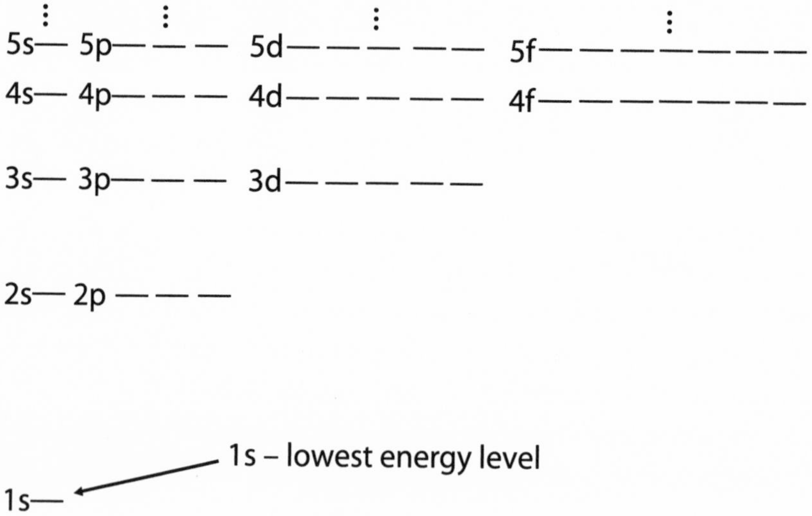

creases, the number of different states (orbitals) associated with the particular n increases. Hydrogen is a special case because it only has one electron. For hydrogen, all orbitals with the same n have the same energy. As discussed in the next chapter, for atoms with more than one electron, for a given n, orbitals with different l values have different energies.

HYDROGEN ATOM s ORBITALS

Although the hydrogen energies only depend on the principal quantum number n, l and m still play an important role. These quantum numbers determine the shapes of the orbitals, and they determine other aspects of the hydrogen atom's properties. For example, the m quantum number is called the magnetic quantum number. The three 2p orbitals, $2p_1$, $2p_0$, and $2p_{-1}$, differ by their m quantum number. When the hydrogen atom is put in a magnetic field, the energies of these three orbitals are no longer the same.

From the energy levels calculated with the Schrödinger equation (see Figure 10.1) it is clear how the empirical diagram in Figure 9.3 arises. The optical transitions seen in the line spectrum of the hydrogen atom and described by the Rydberg formula are transitions between the energy levels of the hydrogen atom, energy levels that are calculated using quantum theory with no adjustable parameters.

As mentioned above, the n, l, and m quantum numbers all go into determining the shapes of the wavefunctions. The s orbitals have $l = 0$. $l = 0$ means that the electron has no angular momentum in its motion relative to the nucleus of the atom. All directions look the same, so s orbitals are spherically symmetric three-dimensional probability amplitude waves. Figure 10.2 shows schematic representations of the 1s, 2s, and 3s orbitals (probability amplitude waves). The darker shading indicates a greater probability of finding the electron that distance from the center. The distances at which

FIGURE 10.2. *The 1s, 2s, and 3s orbitals shown in two-dimensional representations. These are actually spherical. Darker represents a greater probability of finding the electron. Solid circles are distances with peaks in the probability. Dashed circles are nodes where the probability goes to zero. The way the orbitals are represented, they have a fairly sharp outer edge. The orbitals are waves that become very small at large distances, but only decay to zero as the distance from the center goes to infinity.*

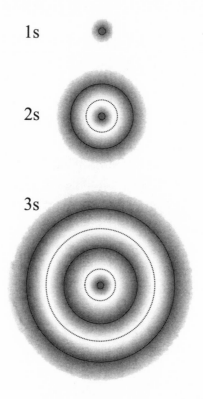

1s

2s

3s

the probabilities have maxima are shown by the solid circles. The centers of the white regions in the 2s and 3s orbitals (dotted circles) are nodes, that is, regions where the probability of finding the electron goes to zero. In going from the 1s to the 2s to the 3s, the size of the orbital becomes much larger. The electron has a greater probability of being found further away from the nucleus as the n quantum number increases.

The increased size of the orbitals is the reason that the energy increases as the quantum number n increases. The formula for the energy levels of the hydrogen atom has a negative sign in front of it, $E_n = -R_H/n^2$. We use a sign convention that a lower energy is a more negative energy. The hydrogen atom is composed of a proton and an electron attracted to each other through a Coulomb interaction, that is, an electrostatic interaction. Opposite charges attract. The proton is positively charged and the electron is negatively charged. When a proton and electron are infinitely far apart, they do not feel each other. There is no attraction because they are so far apart. The system has its zero of energy when the particles are separated at infinity. The electron and proton attraction increases as they get closer to each other. The energy of the system decreases, becoming increasingly negative. The 2s orbital has the electron further away from the proton on average than the 1s, and the 3s orbital has the electron still further away from the proton on average. This is clear from Figure 10.2. As the quantum number increases, the energy is a smaller negative number. For larger values of n, it takes less energy to separate the electron and proton, that is, to ionize the atom. Ionization is the process of pulling the electron out of an atom so that they are no longer bound together. For n = 1, it takes an energy of R_H to ionize the atom. This is the amount of energy that needs to be put into the atom to overcome the binding of $-R_H$. When n = 2, it only takes $R_H/4$ to ionize a hydrogen atom. When n = 3, even less energy, $R_H/9$, is needed to ionize the atom.

SPATIAL DISTRIBUTION OF s ORBITALS

To get a better feel for the spatial distribution of the probability of finding the electron in some position, it is useful to make two types of plots of the wavefunctions. One is just to plot the wavefunction as a function of distance from the nucleus. This type of plot is useful but somewhat misleading. The second type of plot is called a radial distribution function, which will be described shortly. Figure 10.3 is a plot of the wavefunction $\Psi(r)$ as a function of the distance from the proton, which is at the center of the atom. This type of plot is the probability amplitude of finding the electron along a single line radially outward from the center. In Figure 10.2, r is along a horizontal line starting at the center of the shaded electron distribution and moving outward to the right. Figure 10.3 shows that the probability of finding the electron along a single line decreases rapidly, and is close to zero by a distance from the nucleus of 3 Å.

The problem with the type of plot shown in Figure 10.3 is it

FIGURE 10.3. *A plot of the 1s wavefunction* $\Psi(r)$ *as a function of r, the distance from the proton.* $\Psi(r)$ *is proportional to the probability of finding the electron along a line radially outward from the center of the atom. The distance r is in Å, which is* 10^{-10} *m.*

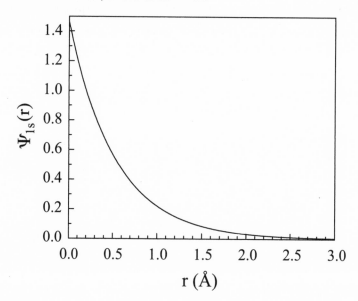

does not account for the three-dimensional nature of the atom. Looking at the 1s orbital in Figure 10.2, you can see that you can find the electron at some distance from the center by moving along a line to the right, but also along a line to the left, or up, or down. You can also move in any diagonal direction a distance r and have the same probability of finding the electron. Since the atom is three dimensional, you can also move in or out of the page and find the electron. If you want to know the probability of finding the electron a certain distance r from the proton, you need to sum all of these different radial directions.

What is really being asked is what is the probability of finding the electron a certain distance from the nucleus when I add together all possible directions? The way to state this question is, what is the probability of finding the electron in a thin spherical shell with the spherical shell having a radius of r? As r gets bigger, the volume of the thin spherical shell increases, which for some distance, offsets the fact that the wavefunction is decreasing. To understand the roll of the thin spherical shell, consider a number of hollow rubber balls each with the same thickness rubber wall. A ball with a small radius (small r) will have less rubber in the wall than a ball with a large radius. If you just went in a single straight line from the center of the ball to the wall, and where you hit the wall you asked what is the thickness of the rubber, it would be independent of the radius of the ball. But it is clear that a large hollow ball has more rubber in its wall then a small ball.

The surface area of a sphere is $4\pi r^2$, where r is the sphere's radius. If you multiply this by the wall thickness, you have the volume of the rubber in the ball. Now it is clear that a larger ball has a lot more rubber in the wall than a small ball. If you double the radius, the amount of rubber increases by a factor of 4. Another important fact is that as r goes to zero, the amount of rubber in the ball goes to zero because the surface area, $4\pi r^2$, goes to zero. Asking if an electron is a distance r from the nucleus is like asking how much rubber is in the wall of a ball of radius r. It is necessary to account for the increasing surface area as the radius increases.

THE RADIAL DISTRIBUTION FUNCTION

The radial distribution function is exactly what we need to take into account the three-dimensional nature of an atom. As r is increased and we look in all directions to find the electron, we must include a factor of $4\pi r^2$. The radial distribution function is a plot of the probability of finding the electron a distance r from the nucleus for all directions. As discussed in Chapter 5, the Born interpretation of the wavefunction says that the probability of finding a particle in some region of space is proportional to the absolute value squared of the wavefunction. Here we want the probability of finding the electron in a thin spherical shell that has radius r. This is the radial distribution function, which is given by $4\pi r^2|\Psi|^2$. The vertical lines mean absolute value. For the functions we are dealing with, we just need to square the wavefunction.

Figure 10.4 displays the radial distribution function for the 1s state of the hydrogen atom. The distance that has the maximum probability is not the center of the atom because the volume of the spherical shell goes to zero as r goes to zero. The vertical line shows the location of the maximum in the probability distribution. It is r = 0.529 Å. This is an important and interesting number. In Bohr's old quantum theory of the hydrogen atom, the 1s state had the electron going in a circular orbit with a radius of 0.529 Å. This distance is called the Bohr radius and is given the symbol a_0. What we see from the correct quantum mechanical treatment of the hydrogen atom is that the electron is a probability amplitude wave with the distance for the maximum probability equal to the Bohr radius a_0. This is not a coincidence. The Bohr radius is actually a fundamental constant. It is given by

FIGURE 10.4. *A plot of the radial distribution function for the 1s orbital as a function of r, the distance from the proton. The radial distribution function is the probability of finding the electron in a thin spherical shell a distance r from the proton. The radial distribution function takes into account that the electron can be found in any direction radially outward from the proton. The distance r is in Å, which is 10^{-10} m.*

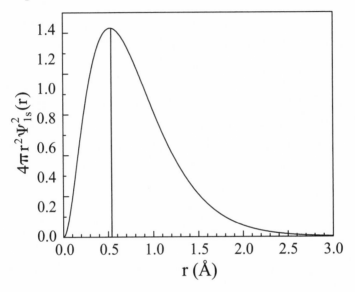

$$a_0 = \frac{\epsilon_o h^2}{\pi \mu e^2}$$

where all of the parameters were given above when the Rydberg constant was defined in terms of fundamental constants. In fact, the energy levels of the hydrogen atom can be written in terms of the Bohr radius as

$$E_n = -\frac{e^2}{8\pi\epsilon_o a_0 n^2}$$

Figures 10.5 and 10.6 show plots of the wavefunctions (top panels) and the radial distribution functions (bottom panels) for the 2s

FIGURE 10.5. *A plot of the 2s hydrogen atom wavefunction (top panel) and the radial distribution function (bottom panel), as functions of r, the distance from the proton. The wavefunction begins positive, goes through a node at slightly more than 1 Å ($2a_0$), and then decays to zero. The radial distribution function shows that the maximum probability of finding the electron peaks at about 2.8 Å, with most of the probability between 2 and 4 Å (see Figure 10.2). The distance r is in Å, which is 10^{-10} m.*

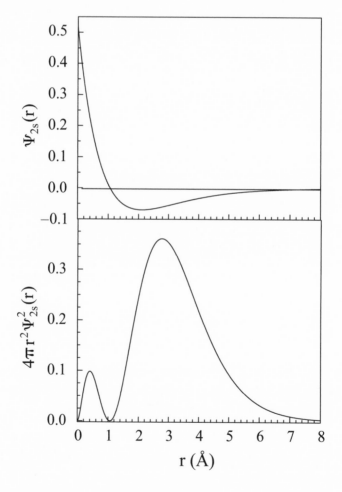

FIGURE 10.6. *A plot of the 3s hydrogen atom wavefunction (top panel) and the radial distribution function (bottom panel), as functions of r, the distance from the proton. The wavefunction begins positive, goes through a node, becomes negative, goes through a second node, and becomes positive again. It then decays to zero. The radial distribution function shows that the maximum probability of finding the electron peaks at about 7 Å, with most of the probability between 5 and 11 Å (see Figure 10.2). The distance r is in Å, which is 10^{-10} m.*

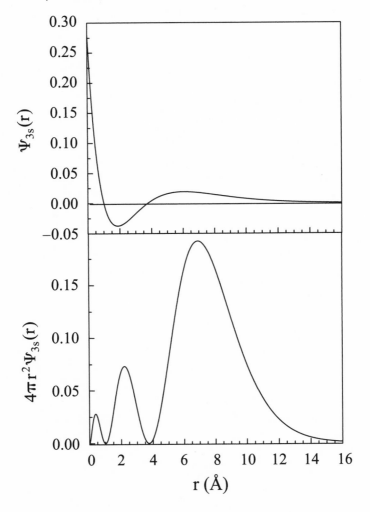

and 3s orbitals. The wavefunction of the 2s orbital has a node, that is, a place where the wavefunction is zero. Nodes were discussed in connection with the particle in a box wavefunctions (see Figure 8.4). At a node the probability of finding a particle, in this case the electron, is zero. The 2s wavefunction begins positive, crosses zero at the node located at twice the Bohr radius, $2a_0$, and then is negative. The wavefunction then decays to zero. By 8 Å the value of the wavefunction is very small. As we have discussed in detail, the wavefunctions are probability amplitude waves. Like other waves, they can be positive or negative. The bottom panel of Figure 10.5 displays the 2s radial distribution function. This is the probability of finding the electron a distance r from the nucleus. Probabilities are always positive because they are the square of the wavefunction, which is always positive. A wave can be positive or negative, but it makes sense that a probability is always a positive number or zero. The radial distribution function shows that most of the probability is between about 2 and 4 Å, which also can be seen in Figure 10.2 but not as quantitatively. The peak of the probability is located at $\sim$2.8 Å.

In Figure 10.6, it can be seen that the 3s wavefunction has two nodes, that is, the wavefunction crosses zero twice. The hydrogen atom wavefunctions have this in common with the particle in a box wavefunctions (see Figure 8.4). For n = 1, there is no node. For n = 2, there is one node. For n = 3, there are two nodes. The number of nodes for the s orbitals is n − 1. The 3s wavefunction begins positive, goes negative, and then becomes positive again. It finally decays to zero, and is very small by about 16 Å. The 3s radial distribution function shows that most of the probability is relatively far from the nucleus. The peak probability is at $\sim$7 Å. Most of the probability is between 5 and 11 Å. The three radial distribution functions shown in Figures 10.4, 10.5, and 10.6 are quantitative plots of the information shown schematically in Figure 10.2. As the principal quantum number, n, gets larger, the s orbitals become larger and have more nodes.

THE SHAPES OF THE p ORBITALS

For the 2s orbital, n = 2, l = 0, and m = 0. However, for n = 2, l can also equal 1 with the associated three values of m, m = 1, 0, −1. The three different m values give rise to the three different 2p orbitals. These are shown in the energy level diagram, Figure 10.1. The three different 2p orbitals are represented schematically in Figure 10.7. As mentioned above, because of their shapes, the 2p orbitals are usually referred to as $2p_z$, $2p_y$, and $2p_x$. Each orbital has two lobes, a positive lobe and a negative lobe. Which lobe is assigned positive and negative is arbitrary, but the sign must change because there is an angular nodal plane. The $2p_z$ orbital has its lobes pointed along the z axis. The nodal plane (shaded in the figure) is the xy plane. This plane (z = 0) is a plane where the probability of finding the electron is zero. The sign of a wavefunction changes when it passes through a node. In the 2s orbital, there is a radial node. A radial node is a distance from the center at which there is a node. Each p orbital has an angular node, that is, directions (a plane) where there is a node. The 2p orbitals do not have a radial node, but the 3p orbitals have a radial node in addition to the angular nodal plane, and the 4p orbitals have two radial nodes, etc.

The $2p_y$ orbital has its lobes pointed along the y axis, and its nodal plane is the xz plane. The $2p_x$ orbital has its lobes pointed along the x axis, and its nodal plane is the yz plane. The schematic illustrations of the 2p orbitals in Figure 10.7 are like the representations of the s orbitals in Figure 10.2. Figure 10.7 gives a feel for the regions that have a large amount of electron probability amplitude. However, it is important to recognize that these are probability amplitude waves that die out smoothly away from the nucleus. In the figure, the lobes terminate abruptly, but the wave-functions decay at long distances in a manner similar to that shown in Figure 10.3 for the 1s orbital. Nonetheless, Figure 10.7 is useful to get a feel for the shapes of the 2p orbitals. These shapes will be very important when we discuss molecular bonding and the shapes of molecules.

THE SHAPES OF THE d ORBITALS

When n = 3, l can equal 0 to give the 3s orbital. l can equal 1 with m = 1, 0, −1 to give the three different 3p orbitals. In addition, l can equal 2 with m = 2, 1, 0, −1, −2 to give five different 3d orbitals. These are shown in the energy level diagram, Figure 10.1. Figure 10.8 shows the five different 3d orbitals. Like the p orbitals,

FIGURE 10.7. *Schematic of the three hydrogen atom 2p orbitals, $2p_z$, $2p_y$, and $2p_x$. Each orbital has two lobes, one positive and one negative. Each has an angular nodal plane, that is, a plane where the probability of finding the electron is zero. The $2p_z$ orbital has its lobes along the z axis, and the nodal plane is the xy plane, which is shaded. The $2p_y$ orbital has its lobes along the y axis, and the nodal plane is the xz plane. The $2p_x$ orbital has its lobes along the x axis, and the nodal plane is the yz plane. The lobes in each diagram show where most of the electron probability amplitude is located. These probability amplitude waves delay smoothly to zero away from the nucleus (the proton) and do not stop abruptly as in the diagrams.*

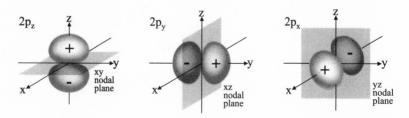

the d orbitals are often given names that reflect their shapes rather than labeling them with the quantum number, m. Four of the orbitals have the same basic shape. Each has four lobes and two angular nodal planes. Two of the lobes are positive and two are negative. When a nodal plane is crossed the wavefunction changes sign. The fifth orbital, the d_{z^2}, has a different shape, but it still has two angular nodal surfaces. One is the xy plane, and the other is the conical surface shown in the diagram. Like the p orbitals, the shaded regions in Figure 10.8 indicate where most of the electron probability amplitude is found. These probability amplitude waves go smoothly to zero as the distance from the nucleus increases.

When n = 4, in addition to s, p, and d orbitals, l can equal 3, which gives rise to the seven possible m values. These are the seven f orbitals. The f orbitals have three angular nodes and very complicated shapes. As will be discussed in the next chapter on atoms bigger than hydrogen, only very heavy elements have f orbitals containing electrons, and the f electrons do not usually participate in making chemical bonds. Many molecules, particularly those in which the prime element is carbon called organic molecules, involve mainly 2s and 2p orbitals. However, molecules that contain heavier elements, such as metals, may involve d orbitals as well.

In Chapter 11, we will build on our discussion of the hydrogen atom to understand the properties of all atoms. Because these larger atoms contain more than one electron, the fourth quantum number, s, will come into play. By applying some simple rules, we will be able to understand many of the properties of atoms and how they form molecules.

FIGURE 10.8. *Schematics of the five hydrogen atom 3d orbitals, which are named in relation to their shapes. Each orbital has two angular nodes, as well as positive and negative lobes. The angular nodes are the planes in four of the diagrams and the cones and disk in the fifth diagram. When a nodal plane is crossed, the wavefunction changes sign. The lobes in each diagram show where most of the electron probability amplitude is located. Four of the orbitals consist of four lobes. The d_{z^2} orbital has a different shape. It sill has two nodal surfaces, the xy plane and the conical surfaces. These probability amplitude waves decay smoothly to zero away from the nucleus (the proton) and do not stop abruptly as in the diagrams.*

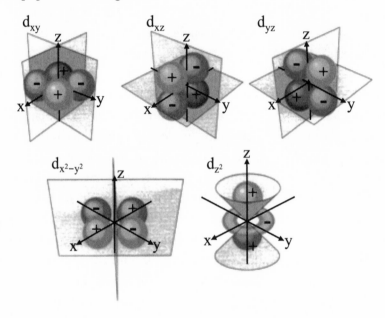

11

Many Electron Atoms and the Periodic Table of Elements

THE PROPERTIES OF ATOMIC and molecular matter are determined by the quantum mechanical details of the atoms that make up a substance. Common table salt is sodium chloride, NaCl. Na is the symbol for the sodium atom. Sodium's atomic number is 11. The atomic number is the number of protons in the nucleus, that is, it is the positive charge on the nucleus. Na has 11 protons in its nucleus and 11 negatively charged electrons. Chlorine (symbol Cl) has an atomic number of 17. Cl has 17 protons in its nucleus and 17 electrons. When table salt, which is composed of little white crystals of NaCl, is put in water, it dissolves. In water, the Na becomes the positively charged sodium ion, Na^+ (sodium has lost an electron), and the Cl becomes the negatively charged chloride ion, Cl^- (chlorine has gained an electron). Sodium gives up an electron to chlorine to form the sodium cation (positively charged ion) and the chloride anion (negatively charged ion). The charges on the sodium cation and the chloride anion make these ions very soluble in water.

Methane is natural gas that we burn in our gas stoves, gas clothes dryers, and power plants. Its chemical formula is CH_4. This means it is composed of one carbon atom (symbol C, atomic number 6) bonded to (attached to) four hydrogen atoms (symbol H, atomic number 1). Methane does not become ions when put in water. In fact, it does not dissolve in water. Unless it is raised to very high temperature as in a flame, it does not come apart at all. Why does NaCl become separated ions, Na^+ and Cl^-, when dissolved in water, while carbon always makes four chemical bonds and methane does not come apart in water to form ions? The answers to these questions and many of the properties of all of the atoms can be understood by looking at the nature of many electron atoms and the collection of systemic information about atoms contained in the Periodic Table of the Elements.

HYDROGEN IS DIFFERENT

The hydrogen atom is different from all other atoms, and the difference is highly significant. The hydrogen atom consists of a positively charged nucleus (a proton) and one negatively charged electron. The sole electrostatic interaction is the attraction of the electron to the proton because oppositely charged particles attract. The next simplest atom is helium. Helium consists of a positively charged nucleus, with a positive charge of $+2$ (symbol He, atomic number 2) and two electrons, each with a negative charge of -1. Each electron is attracted to the nucleus. In addition, the two electrons repel each other because like changes repel. The repulsion is referred to as electron-electron repulsion. Because a hydrogen atom has only a single electron, there is no electron-electron repulsion.

In the energy level diagram for the hydrogen atom (Figure 10.1), all of the orbitals with the same principal quantum number n have the same energy. So the 2s and the 2p orbitals have the same energy. The 3s, 3p, and 3d orbitals all have the same energy, and so on. The fact that energy only depends on the principal quantum number is a result of hydrogen having a single electron. In Figures

10.2, 10.7, and 10.8, the s, p, and d orbitals have very different shapes. However, in hydrogen, the electron, on average, is the same distance from the nucleus independent of the shape of the orbitals. So an electron has the same energy whether it is in a 3s, 3p, or 3d orbital. Why? Because the electron has the same attraction to the nucleus when averaged over the spatial distribution given by its 3s, 3p, or 3d wavefunctions.

THE ORBITAL SHAPES MATTER FOR ATOMS LARGER THAN HYDROGEN

With more than one electron, the shapes of the orbitals matter. In helium, if its two electrons are placed in the 2s orbital the energy is lower than if they are placed in a 2p orbital. On average, two electrons in the 2s orbital are farther apart than two electrons in a 2p orbital. Electron-electron repulsion increases the energy. Because the two electrons are farther apart in the 2s orbital, the electron-electron repulsion (increase in energy) is not as severe as having the two electrons in a 2p orbital. Therefore, the 2s orbital in many electron atoms (all atoms but hydrogen) is lower in energy than the 2p orbital. For n = 3, two electrons on average are farther apart in a 3s orbital than in a 3p orbital, and two electrons in a 3p orbital are farther apart than if they are in a 3d orbital. So, the 3s orbital is lower in energy than the 3p orbitals, which are lower in energy than the 3d orbitals. However, the 3s orbital is higher in energy than the 2s orbital. On average electrons in a 3s orbital are farther from the nucleus because the 3s orbital is larger than the 2s orbital (see Figures 10.2, 10.5, and 10.6), and therefore have a weaker attractive interaction with the nucleus. Less attractive interaction results in a higher energy. The attraction to the nucleus binds the electron to the nucleus. The sign convention is that the stronger the binding, the lower the energy. The electron falls into the attractive well of the positively charged nucleus. The stronger the attraction, the deeper the electron is in the well. It will take more energy to remove the electron from the well, that is, pull it away from the nucleus.

MANY ELECTRON ATOM ENERGY LEVELS

For a given principal quantum number n, the order of the energy is ns < np < nd < nf. For the same type of orbital, the larger the n the higher the energy. The important feature of many electron atoms is that the energy depends on two quantum numbers, n and l. l is the angular momentum quantum number that determines the shape of the orbital. Figure 11.1 is the energy level diagram for many electron atoms. For n = 1, there is only one type of orbital, l = 0, an s orbital. So the 1s orbital has the lowest energy level. For n = 2, l can be 0 or 1. The l values give rise to the 2s orbital and the three different 2p orbitals. With l = 1, there are three possible values of m, m = 1, 0, −1. This is the same as in hydrogen. The big differ-ence is that for many electron atoms, the 2s orbital is lower in en-ergy than the 2p orbitals, as shown in Figure 11.1. For n = 3, there is the 3s orbital, the 3p orbitals, and the 3d orbitals. As can be seen in Figure 11.1, the 3s orbital is below (lower in energy) the 3p orbit-als, which are below the 3d orbitals.

A very important aspect of the ordering of the energy levels is that energy levels with different n quantum numbers are inter-spersed. Although the 3d orbitals are above the 3p orbitals, the 4s orbital energy is actually below that of the 3d orbitals (see Figure 11.1). The ordering of the orbitals is also shown in Figure 11.1. We see that the energy levels go 1s, 2s, 2p, 3s, 3p, 4s, 3d, 4p, 5s, 4d, and so on. As discussed below, switching the order between the 4s and the 3d gives rise to what is called the first transition series, and switching the 5s and the 4d gives rise to the second transition series. The ordering is very important in determining the properties of various atoms. The switches in the order and the meaning of the transition series will become clear as we discuss the Periodic Table

FIGURE 11.1. *Energy level diagram for atoms with many electrons. The spacings between the levels are not to scale. The energy depends on the principal quantum number, n, and the angular momentum quantum number, l, in contrast to the hydrogen atom (Figure 10.1), where the energy only depends on n. For n = 4, there is a single s orbital (l = 0), three different p orbitals (l = 1), five different d orbitals (l = 2), and seven different f orbitals (l = 3).*

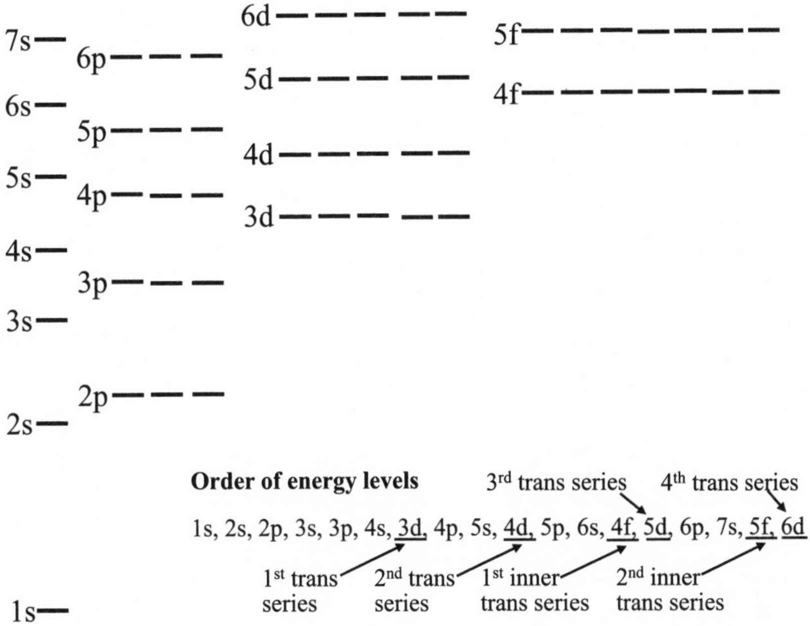

of the Elements. However, first we need to discuss how to "put" the electrons into the energy levels shown in Figure 11.1.

THE THREE RULES FOR PUTTING ELECTRONS IN ENERGY LEVELS

The hydrogen atom has a nucleus with charge +1 and a single negative electron. The helium atom has a nucleus with charge +2 and two negative electrons. Next comes the lithium atom (symbol Li) with a +3 charge on the nucleus (atomic number 3) and three negative electrons, which is followed by beryllium (Be) with a +4 nucleus and four negative electrons, and so on. The question is if we have an atom with a certain number of electrons, like beryllium with four, which energy levels do these four electrons go into? For hydrogen, the lowest energy state is the one in which the single electron is in the 1s orbital. If we excited the hydrogen 1s electron to say a 2p state (adding energy by absorbing light or with an electrical arc), it will fall back to the lowest energy state and conserve energy by emitting a photon. Such photon emission from the various energy levels of the hydrogen atom gives rise to its line spectrum discussed in Chapters 9 and 10. But it is not clear what to do when there is more than one electron. Should the four electrons of beryllium all go in the 1s orbital? It turns out that this is impossible.

Quantum theory, confirmed by countless experiments, has given us three rules that tell us how to place the electrons into the energy levels (Figure 11.1) to obtain the configuration of electrons for the various atoms. We use what is called the Aufbau procedure, that is, the build-up procedure. The three rules tell us how to place the electrons in the energy levels in the correct order to represent atoms. We will build up the atoms and construct the periodic table by "putting" more and more electrons for bigger and bigger atoms into the proper energy levels. Many properties of atoms, their tendency to gain or lose electrons to form ions, and the number of chemical bounds they form will be understandable from this Aufbau procedure, which gives rise to the form of the Periodic Table.

Rule 1—the Pauli Exclusion Principle

Rule 1 is the Pauli Exclusion Principle. It states that no two electrons in an atom (or molecule) can have all four quantum numbers identical. There are four quantum numbers, n, l, m, and s. For hydrogen we only used the first three, but now s is important. s can only have two values, s = +1/2 or −1/2. Therefore, a given orbital defined by the quantum numbers n, l, m can have at most two electrons in it. One of the electrons will have s = +1/2 and one will have s = −1/2. For example, the 1s orbital has n = 1, l = 0, m = 0, and s = +1/2 or −1/2. Therefore, two electrons can go into the 1s orbital, one with "spin" +1/2 and one with spin −1/2. For the 2p orbitals, n = 2, l = 1, m = 1, 0, −1 and s = +1/2 or −1/2. The orbitals are p_x, p_y, and p_z (see Figure 10.7). Each of these can have two electrons in it, one with s = +1/2 and the other must have s = −1/2. Therefore, there can be a total of six 2p electrons, two in each of the three orbitals. 3d orbitals have quantum numbers n = 3, l = 2, m = 2, 1, 0, −1, −2, and s = +1/2 or −1/2. There are five 3d orbitals, and two electrons can go in each (s = +1/2 or −1/2) for a total of 10 d electrons, two in each of five orbitals. Finally, there are seven 4f orbitals with quantum numbers, n = 4, l = 3, m = 3, 2, 1, 0, −1, −2, −3, and s = +1/2 or −1/2. The result is a total of 14 f electrons, two in each of seven orbitals.

When two electrons are in a single orbital, the spins are said to be paired. An electron in an orbital (energy level) is represented by an arrow (see Figure 11.2). The spin quantum number s = +1/2 is represented by an arrow pointing up. The spin quantum number s = −1/2 is represented by an arrow pointing down. In any single orbital there can be at most one up arrow and one down arrow.

Rule 2—Lowest Energy First but Don't Violate the Pauli Principle

Rule 2 is the orbitals are filled with electrons in order of increasing energy. Electrons are placed in the lowest possible energy level first, but the Pauli Principle cannot be violated. So for the helium atom (He), both electrons can go in the 1s energy level, one with spin up (s = +1/2) and one with spin down (s = −1/2). Three of the

FIGURE 11.2. *Left-hand side: an electron represented by an arrow in an orbital. Right-hand side: two electrons in the same orbital. The s quantum numbers must be + 1/2 and − 1/2, represented by an up arrow and a down arrow, to obey the Pauli Exclusion principle. The spins are said to be paired.*

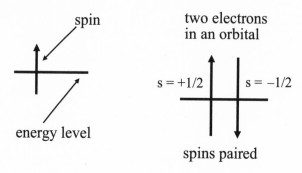

quantum numbers are the same, but s is different, so the Pauli Principle is not violated. Li is the next largest atom with three electrons. The third electron cannot go in the 1s level because it would have all four quantum numbers, n, *l*, m, and s, the same as one of the other two electrons. So, the third electron must go into a higher energy level, the 2s orbital. The 2s is the lowest possible level for the third electron. Therefore, rule 2 dictates it will go there.

Rule 3—Hund's Rule: Don't Pair Spins If Possible Without Violating Rules 1 and 2

Rule 3 is called Hund's Rule. Hund's Rule states that electrons remain unpaired if possible when filling orbitals of identical energy. Figure 11.3 illustrates Hund's Rule using the 2p orbitals as an example. The first electron, labeled 1 in the figure, is placed in the $2p_x$ orbital. This choice is arbitrary since all three 2p orbitals have the same energy. According to Hund's Rule, the second electron will go into one of the other two 2p orbitals, which have the same en-

FIGURE 11.3. *Illustration of Hund's Rule. When filling the 2p orbitals, electron 1 is placed in 2p$_x$, electron 2 in 2p$_y$, and electron 3 in 2p$_z$. These are all spin up. Electron 4 will have to have its spin down, that is, pair, to avoid violating the Pauli Principle.*

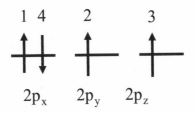

ergy, so that the spins are not paired. Here it is placed in the 2p$_y$ orbital. The third electron must go in the 2p$_z$ orbital, which is the only choice that obeys Hund's Rule as well as Rules 1 and 2. Finally, the fourth electron pairs with one of the other electrons. As shown, it is in the 2p$_x$. It must be spin down to obey the Pauli Principle, Rule 1.

Hund's Rule comes about because it gives electron configurations with the lowest possible energy. Putting two electrons in two different 2p orbitals keeps them further apart on average than putting them in the same orbital. The energy is lowered because keeping the electrons further apart reduces electron-electron repulsion. So what Hund's Rule says in effect is to put electrons in different orbitals if possible. While the energy reduction associated with keeping the electrons unpaired is significant, the amount the energy is lowered is not large. Therefore, it is better to pair electron 4 in the 2p$_x$ than to put it unpaired in the next higher energy orbital, the 3s.

THE PERIODIC TABLE OF ELEMENTS

We have laid down the rules for putting electrons in the energy levels shown in Figure 11.1. Now we will use these rules to understand many properties of atoms and the Periodic Table of Elements. In addition, these same rules will be very important in our discussion of molecules presented in subsequent chapters. But first we need to introduce the Periodic Table shown on the next page.

The Periodic Table has one box for each element. The box has the symbol for the element, as well as its atomic number. The atomic number is the number of positively charged protons in the nucleus. For a neutral atom (not a positively or negatively charged ion), the atomic number is also the number of negatively charged electrons. The form of the Periodic Table will be explained in detail below. In the upper left corner is hydrogen, with symbol H and atomic number 1. In the upper right corner is helium, with symbol He and atomic number 2. Below hydrogen is lithium, with symbol Li and atomic number 3. Many of the symbols are just abbreviations for the names. But this is not always true. For example, lead is element 82 and has the symbol, Pb. Pb derives from the Latin word for lead, *plumbum*. Because it is not always obvious from the symbol what the name of the element is, Table 11.1 gives the element names, their symbols, and their atomic numbers. The table is alphabetical by the element name. If you have the symbol but don't know the name, move down the symbol column until you find the one you want.

The Periodic Table (Figure 11.4) is coded for metals (white), semimetals (semiconductors, dark gray), and nonmetals (light gray). The semimetals are a band between the metals (most of the elements) and the nonmetals, which are in the upper right portion of the table. At the bottom of the table are two strips of elements called the Lanthanide series and the Actinide series. The Lanthanide series, usually referred to as the Lanthanides, begins with the element La, lanthanum, and the Actinides begins with the element Ac, actinium. These two strips go in the gap indicated in the table. These two series of atoms, which involve the f orbitals, are placed below the rest so that the table will not be too wide.

FIGURE 11.4. *The Periodic Table of Elements.*

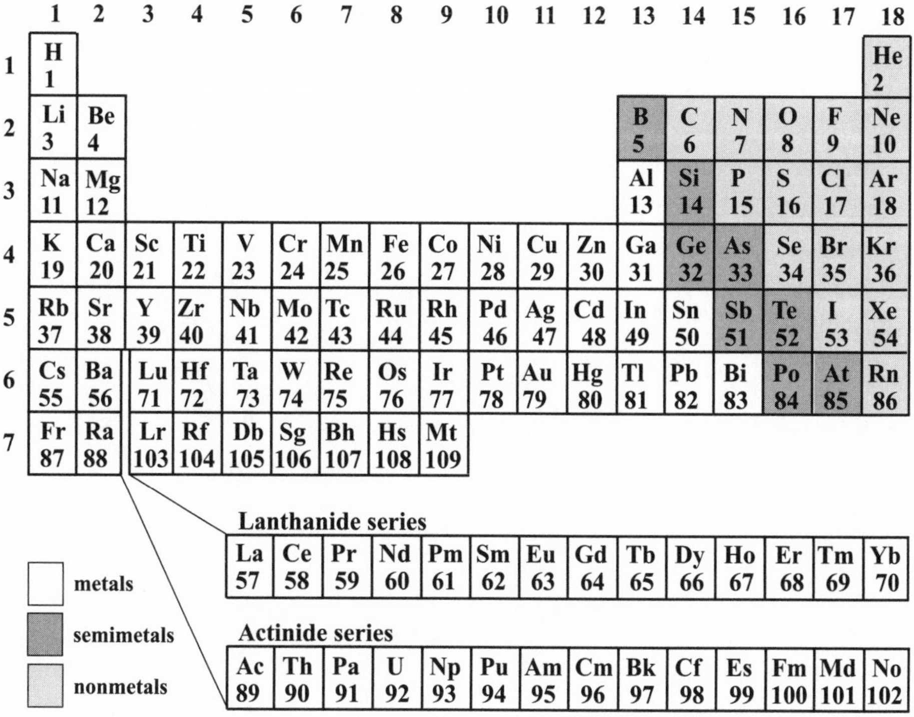

Before reviewing the properties of the elements, we will quickly go through the first two rows of the Periodic Table to get a feel for the layout and to understand what a "closed shell" electron configuration is. Then we will come back and use the table to understand properties of the elements.

The Periodic Table Layout

Referring to the energy level diagram (Figure 11.1) and the three rules for placing electrons in the energy levels, hydrogen (symbol H, atomic number 1) has one electron in the 1s orbital from the rule, lowest energy first without violating the Pauli Principle. H is

TABLE 11.1. *List of the Elements—Alphabetical by Element*

Element	Symbol	Atomic Number	Element	Symbol	Atomic Number
Actinium	Ac	89	Mercury	Hg	80
Aluminum	Al	13	Meitnerium	Mt	109
Americium	Am	95	Molybdenum	Mo	42
Antimony	Sb	51	Neodymium	Nd	60
Argon	Ar	18	Neon	Ne	10
Arsenic	As	33	Neptunium	Np	93
Astatine	At	85	Nickel	Ni	28
Barium	Ba	56	Niobium	Nb	41
Berkelium	Bk	97	Nitrogen	N	7
Beryllium	Be	4	Nobelium	No	102
Bismuth	Bi	83	Osmium	Os	76
Bohrium	Bh	107	Oxygen	O	8
Boron	B	5	Palladium	Pd	46
Bromine	Br	35	Phosphorus	P	15
Cadmium	Cd	48	Platinum	Pt	78
Calcium	Ca	20	Plutonium	Pu	94
Californium	Cf	98	Polonium	Po	84
Carbon	C	6	Potassium	K	19
Cerium	Ce	58	Praseodymium	Pr	59
Cesium	Cs	55	Promethium	Pm	61
Chlorine	Cl	17	Protactinium	Pa	91
Chromium	Cr	24	Radium	Ra	88
Cobalt	Co	27	Radon	Rn	86
Copper	Cu	29	Rhenium	Re	75
Curium	Cm	96	Rhodium	Rh	45
Dubnium	Db	105	Rubidium	Rb	37
Dysprosium	Dy	66	Ruthenium	Ru	44
Einsteinium	Es	99	Rutherfordium	Rf	104
Erbium	Er	68	Samarium	Sm	62
Euorpium	Eu	63	Scandium	Sc	21
Fermium	Fm	100	Seaborgium	Sg	106
Fluorine	F	9	Selenium	Se	34
Francium	Fr	87	Silicon	Si	14
Gadolinium	Gd	64	Silver	Ag	47
Gallium	Ga	31	Sodium	Na	11
Germanium	Ge	32	Strontium	Sr	38
Gold	Au	79	Sulfur	S	16
Hafnium	Hf	72	Tantalum	Ta	73
Hassium	Hs	108	Technetium	Tc	43
Helium	He	2	Tellurium	Te	52
Holmium	Ho	67	Terbium	Tb	65
Htdrogen	H	1	Thallium	Tl	81
Indium	In	49	Thorium	Th	90
Iodine	I	53	Thulium	Tm	69
Iridium	Ir	77	Tin	Sn	50
Iron	Fe	26	Titanium	Ti	22
Krypton	Kr	36	Tungsten	W	74
Lanthanum	La	57	Uranium	U	92
Lawrencium	Lr	103	Vanadium	V	23
Lead	Pb	82	Xenon	Xe	54
Lithium	Li	5	Ytterbium	Yb	70
Lutetium	Lu	71	Yttrium	Y	39
Magnesium	Mg	12	Zinc	Zn	30
Manganese	Mn	25	Zirconium	Zr	40
Mendelevium	Md	101			

in the upper left corner of the Periodic Table. It is the first element in row 1. The next element is helium (He, 2). It has two electrons in the 1s orbital with opposite spins (arrow up and arrow down, as shown in Figure 11.2). This configuration obeys the Pauli Principle and the lowest energy first rule, which overrides Hund's Rule because it would take too much energy to put the second helium electron in the 2s energy level (see Figure 11.1). He is in the upper right corner of the Periodic Table; it completes the first row. The first row has the two elements, H and He, with electrons in the $n = 1$ level. The rows are also referred to as shells. Helium completes the first shell. We say that He has a closed shell configuration because it is the largest element with $n = 1$.

The next element is lithium (Li, 3). It has three electrons. The first two electrons go into the 1s energy level obeying the lowest energy first rule. The third electron cannot go into the 1s orbital because that would violate the Pauli Principle. So the third electron goes in the 2s orbital. Li is below H in the Periodic Table. H is the first element with an electron in the $n = 1$ shell. Li is the first element in the second row, the $n = 2$ shell. The next element is beryllium (Be, 4). The fourth electron also goes into the 2s orbital. This is the lowest energy state, and it does not violate the Pauli Principle. The next element is boron (B, 5) with five electrons. The fifth electron cannot go in the 2s orbital because that would violate the Pauli Principle, which states that no more than two electrons can go into a single orbital and those two must have opposite spins (spin quantum number, $s = +1/2$ and $s = -1/2$). So, the fifth electron goes into a 2p orbital. It doesn't matter which 2p orbital. Following Figure 11.3, we will put it in the $2p_x$ orbital. There is a gap in the table between Be and B. The reason for this will be clear when we discuss the fourth row below. The next element is carbon (C, 6) with six electrons. Now Hund's Rule comes into play, and we put the sixth electron into the $2p_y$ orbital following the layout of Figure 11.3. Nitrogen is next (N, 7). Following Hund's Rule, the seventh electron of N goes into the $2p_z$ orbital so none of the electrons in the p orbitals are paired. Oxygen (O, 8) has eight electrons. The eighth electron must pair because the first seven electrons put two electrons in the 1s, two electrons in the 2s, and one electron in each of the 2p orbitals. To avoid spin pairing requires putting the eighth electron in the 3s orbital, which is much higher energy. So as in Figure 11.3, the eighth electron goes in the $2p_x$ orbital. Fluorine (F, 9), has its ninth electron go into the $2p_y$ orbital. Finally, Neon (Ne, 10) completes the n = 2 row or shell with 10 electrons. The 10th electron goes in the $2p_z$ orbital.

Closed Shell Configurations

The electron configuration for neon is shown in Figure 11.5. No additional electrons can go in the second shell (n = 2 orbitals) without violating the Pauli Principle. As will be discussed, the elements, He, Ne, Ar, Kr, etc., that run down the last column on the right-hand side of the Periodic Table are special. These elements are called the noble gases. They all have closed shells, that is, the next

FIGURE 11.5. *The electron configuration for neon (Ne, 10). The second shell is complete.*

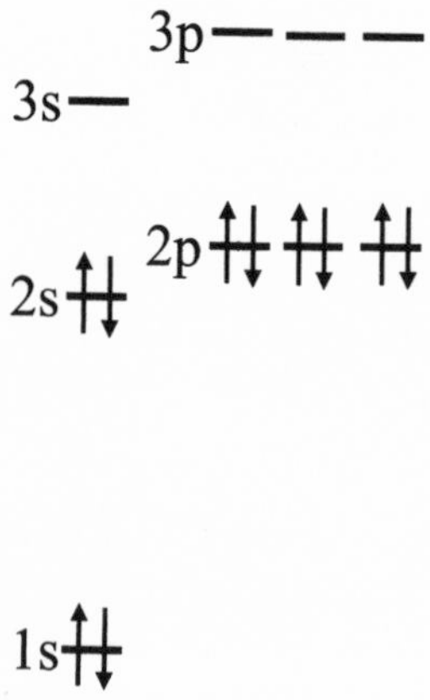

element with one more electron goes into an orbital with the n quantum number one unit larger, which is substantially higher in energy.

Atoms Want to Form Closed Shell Configurations

We are now ready to use the energy level diagram, Figure 11.1, and the three rules for placing electrons in the energy levels to understand the structure of the Periodic Table and the properties of the atomic elements. The following chapters investigate in considerable detail what holds atoms together to make molecules. However, a great deal can be learned from an amazingly simple rule: *Atoms will gain or lose electrons to obtain the nearest closed shell configuration.* The closed shell electron configurations are the configurations of the noble gases that comprise the right-hand column of the Periodic Table. A closed shell configuration is particularly stable. The noble gases, also called the inert gases, have the closed shell configuration and are essentially chemically inert. The noble gases with small atomic numbers, helium, neon, and argon, do not form chemical compounds at all. The higher atomic number noble gases can be forced to form a small number of compounds under specialized conditions. Atoms other than the noble gases change in ways so that they achieve a stable closed shell electron configuration.

There are two ways that an atom can change the number of electrons it has to achieve a closed shell configuration. The first is to become a positive ion (cation) or negative ion (anion). The atom gives up one or more electrons and becomes positively charged (cation), or the atom takes on extra electrons, and becomes negatively charged (anion). The other is for an atom to share electrons with one or more other atoms. When two or more atoms share electrons, it is as if each atom has the electrons it needs. So an atom with fewer electrons than the number needed for the next closed shell configuration obtains the correct number, but so do the other atoms that are involved in the sharing. When atoms share electrons to get to the next closed shell number of electrons, the sharing holds the atoms together. The sharing makes the energy of the combined atoms lower than the energies of the individual open shell atoms. The lowering of the energy bonds the atoms together. This type of chemical bond is called a covalent bond. Covalent bonds are responsible for most of chemistry. The detailed nature of the covalent bond is presented for the simplest molecule, the hydrogen molecule in Chapter 12, and more complex molecules are discussed in subsequent chapters.

The Properties of Atoms

To begin our discussion of the properties of atoms based on the Periodic Table, we start with hydrogen. As usual, hydrogen is special because it only has one electron and is the first element in the Periodic Table. For the first row of the periodic table, helium has a closed shell with two electrons in the 1s orbital. Hydrogen can obtain the helium closed shell configuration by sharing with another atom to pick up an electron. For example, one H atom can share electrons with a second H atom, to form the hydrogen molecule. The symbol for the hydrogen molecule is H_2. The subscript tells how many of a given type of atom are in a molecule. Because of the sharing, each H atom feels as if it has two electrons, the helium closed shell configuration. As we will see, hydrogen can form other molecules, but because it only needs one electron to get to the helium closed shell configuration, it only makes one chemical bond. Helium has the closed shell. It does not make any chemical bonds. There are no molecules with a helium atom in them. Exactly why this is true is described in detail in Chapter 12. Helium completes the first shell.

The next atom is lithium, Li, which is directly below H on the periodic table. Li can obtain the helium closed shell configuration by giving up an electron. Therefore, Li forms $+1$ ions, Li^{+1}. A solid piece of Li is a metal. Metals can conduct electricity, which means electrons can move easily from one atom to another. The nature of metals and electrical conductivity will be discussed in Chapter 19. Metals have the property that as single atoms they can easily give up one or more electrons. The electron Li loses has to go somewhere. It will go to another atom that wants an electron to form a negative ion. So to form Li^{+1}, Li needs a partner (see the discussion below when we get to the other side of the Periodic Table). The next element is beryllium. Beryllium will give up two electrons to go back to the He closed shell configuration. Therefore, Be will form $+2$ ions, Be^{+2}. Because Be can readily give up electrons, solid beryllium is a metal. The next element is boron. Boron can lose three electrons to go back to the He closed shell configuration. Therefore, it forms $+3$ ions, and it is a metal.

Now things change. The next element is carbon. Carbon would have to give up four electrons to go back to the He configuration, but it can also gain four electrons to go to the next closed shell configuration, that of neon. As shown in Figure 11.5, Ne has a closed shell configuration, with the second shell filled. It has two electrons in the 1s orbital and then has the $n = 2$ shell filled with two electrons in the 2s and six electrons in the three 2p orbitals. Rather than losing so many electrons to go back to the He configuration, C goes forward to the Ne configuration, picking up four electrons by making four covalent bonds. For example, methane (natural gas) is the molecule CH_4. Each H is bonded to the central C. The carbon shares four electrons, one from each of the hydrogens, so C obtains the Ne closed shell configuration by sharing four electrons with the four H atoms. Each H shares one electron with the C, so each H picks up one electron to have the He closed shell configuration. This is very important. Through covalent bonds (electrons sharing), each atom obtains a closed shell configuration. Another exceedingly important fact is that C always makes four bonds because it needs to share four electrons to get to the Ne configuration. This fact is fundamental to organic chemistry and to the chemistry of life. The bonding and chemistry of carbon will be discussed extensively in subsequent chapters

The next element is nitrogen. N needs three electrons to get to the Ne configuration, so it makes three covalent bonds. For example, it will combine with H to make NH_3, which is ammonium. Oxygen needs two electrons to get to the Ne closed shell configuration, so it forms two bonds and, for example, makes H_2O, water. So from these simple considerations we can understand the sequence of compounds, CH_4, NH_3, and H_2O. Bonding involving C, N, and O will be discussed in later chapters to understand molecules involving these atoms, but they will always make 4, 3, and 2 bonds, respectively.

The next atom is fluorine. Fluorine is only one electron away from the Ne closed shell. It only needs one electron to obtain the Ne configuration. It has such a strong affinity for an electron that it tends to form the negative ion F^{-1} by picking up an electron. The electron has to come from somewhere, and F forms what are generically called salts. For example, LiF is a white crystalline solid. In the crystal, an Li that wants to give up an electron to obtain the He configuration gives its electron to an F. The result is that an LiF crystal is composed of Li^{+1} ions and F^{-1} ions. The Li^{+1} ions have the He closed shell configuration, and the F^{-1} ions have the Ne closed shell configuration. LiF crystals, like all salts, dissolve readily in water. The crystal is held together by electrostatic interactions. The positive and negative ions attract each other. They are arranged in the crystal in such a way that the attractive interactions among cations and anions overcome the repulsive interactions of the cations with other cations and the anions with other anions. Water can surround both positive and negative ions in a way that makes the energy of the total system, water surrounding cations and anions, lower than that of an LiF crystal sitting in water. This is called solvation. Water can solvate ions, so ionic crystals like LiF dissolve in water. Solvation is discussed in Chapter 15.

F will form salts with atoms on the left side of the Periodic Table, which want to give up electrons to obtain a closed shell configuration. In LiF, F gains an electron and Li loses an electron. F can also obtain the closed shell Ne configuration by making covalent bonds under some circumstances. As discussed below, it can combine with sulfur (S) to form the compound, SF_2.

After F comes Ne in the periodic table. It has a closed shell (see Figure 11.5). Ne does not want to gain or lose electrons. It does not form chemical compounds. Ne completes the second row of the Periodic Table. After Ne is sodium. Sodium is one electron (3s) past the Ne configuration. Like Li directly above it, Na will readily give up an electron to form the cation, Na^{+1}. It does this to obtain the Ne configuration. Solid sodium is a metal that can conduct electricity (electrons) because its 3s electron is not tightly bound. Like LiF, NaF is a salt that will readily dissolve in water. Next comes magnesium. Mg will give up two electrons to obtain the Ne closed shell configuration, forming Mg^{+2} ions. It is a metal that conducts electricity because it can readily give up its two 2s electrons. It will make salts of the form, for example, MgF_2. This means the crystal has two fluorine anions for every magnesium $+2$ cation. MgF_2 will readily dissolve in water. After Mg comes aluminum. Solid Al is a metal. Al will form Al^{+3} cations.

As with carbon in the second row, things change with silicon. Si will make four covalent bonds to share (effectively gain) four electrons to obtain the argon closed shell configuration (see the Periodic Table). For example, it will form SiH_4. Phosphorous will make three covalent bonds to obtain the Ar configuration, for example, PH_3, and S will make two covalent bonds to get to the Ar closed shell configuration. It will form the compound H_2S, which is hydrogen sulfide, the very smelly and poisonous gas that gives rotten eggs their smell. As mentioned above, S can also make covalent bonds with

F, to form SF_2. After S comes chlorine. Like F, which only needs one electron to obtain the Ne closed shell configuration, Cl only needs one electron to obtain the Ar closed shell configuration, so it tends to form Cl^{-1} by gaining an electron. All of the elements in the column next to the noble gases, the second column from the right in the Periodic Table, form -1 anions. These elements (F, Cl, Br, I, At) are called the halogens. This brings us to common table salt, sodium chloride, NaCl, which is a crystalline solid composed of Na^{+1} and Cl^{-1}. Like LiF, NaCl can dissolve in water because the cations and anions can be solvated by the H_2O molecules. This is in contrast to methane, CH_4, which does not dissolve in water. The carbon and the hydrogens close their shells by sharing electrons with each other to form covalent bonds. If methane separates into pieces, the atoms will no longer have closed shells. This is different than an NaCl crystal, which can come apart into Na^{+1} and Cl^{-1}. Both the cation and anion have closed shells. Molecules that are only composed of carbon and hydrogen, like oil, gasoline, and methane, are called hydrocarbons. They do not dissolve in water. Hydrocarbons are discussed in Chapters 14, 15, and 16. After Cl is Ar. Argon has a closed shell, as shown in Figure 11.6. It has 18 electrons, two in the 1s, two in the 2s, six in the three 2p orbitals, two in the 3s, and six in the three 3p orbitals. Ar is an inert gas. It does not form chemical compounds.

Going Down a Column, Atoms Get Bigger

Going down a column in the Periodic Table, the atoms get bigger. So, Li is bigger than H, Na is bigger than Li, and so forth. Two considerations explain this. First, the additional electrons go into orbitals with a larger principal quantum number, n. H has a 1s electron, Li has a 2s electron, and Na has a 3s electron. Looking at Figures 10.2 through 10.6, which are for hydrogen, you can see that the 3s wavefunction is much bigger than the 2s, which is much

FIGURE 11.6. *The electron configuration for argon (Ar, 18). The third row is complete.*

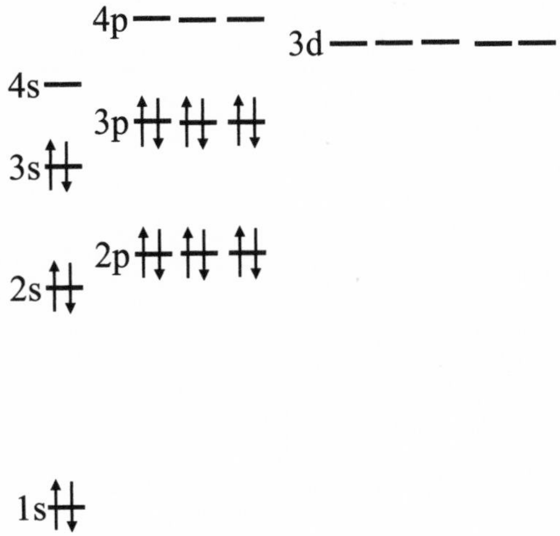

bigger than the 1s. However, as we go down a column, the positive nuclear charge also increases. The nuclear charge is the same as the atomic number, which is given for each atom in the Periodic Table as well as in the List of Elements. The increased nuclear positive charge pulls the negatively charged electrons in closer. This increased attraction is not sufficient to offset the fact that going down a column puts electrons in orbitals with larger principal quantum numbers (n). The increase in size with n wins out over the increased nuclear attraction for the electrons, resulting in larger atoms as you move down a column.

Going Left to Right Across a Row, Atoms Get Smaller

As you go along a row from left to right, the atoms get smaller. So Be is smaller than Li, B is smaller than Be, C is smaller than Be, etc. The reduction in size occurs because all of the atoms have the same principal quantum number n, but the nuclear charge increases. Again two phenomena are playing off against each other. The positive nuclear charge increases as you move to the right along a row. The increased positive charge will pull the electrons in closer to the nucleus. However, there are also more electrons. The negative electrons repel each other (electron-electron repulsion). To reduce the electron-electron repulsion, the electron cloud (probability amplitude wave) gets larger. The positive charge is at the center, pulling all of the electrons in. But the negative electron cloud is spread out around the nucleus. Speaking very classically, at a given instant, the electrons on one side of the atom don't feel (aren't repelled by) the electrons all the way on the other side of the atom as much as they are attracted to the nucleus at the center. So the attraction wins, and the atoms get smaller as you move along a row from left to right.

The First Transition Series

Now we are at the fourth row. After Ar, the first element in the fourth row is potassium, K. K has one 4s electron past the Ar configuration. By now it should be clear that potassium will form K^{+1} ions so that it can obtain the Ar closed shell configuration. Solid K is a metal that conducts electricity. The salt KCl is a small component of sea salt, which is mainly NaCl. KCl dissolves in water to give the ions K^{+1} and Cl^{-1}. Next to K is calcium, Ca. Ca has two 4s electrons past the Ar configuration. It is a metal that forms Ca^{+2} ions, giving up its two 4s electrons to obtain the Ar closed shell configuration. It will form salts like $CaCl_2$, which readily dissolve in water to give a calcium $+2$ cation and two chloride anions.

Now things change again in a big way. The energy level diagram for many electron atoms (Figure 11.1) shows that the 3d orbitals are above the 4s orbitals in energy, but they are below the 4p orbitals. As mentioned earlier in this chapter, the interposition of the 3d orbitals between the 4s and the 4p orbitals gives rise to the first transition series in the Periodic Table. There are five 3d orbitals. The Pauli Principle states that there can be at most two electrons in an orbital. Then there can be 10 electrons in the five 3d orbitals, resulting in the 10 transition metals, scandium through zinc (see the Periodic Table). So after Ca come 10 elements that arise from filling the 3d orbitals. All of these are metals. Many of them are very common metals that we are familiar with in everyday life, such as iron (Fe), copper (Cu), nickel (Ni), zinc (Zn), and chromium (Cr). They can readily form ions. The first two elements in a row, such as K and Ca or Na and Mg, always form cations with a particular charge, $+1$ for the first column (Na^{+1} and K^{+1}) and $+2$ for the second column (Mg^{+2} and Ca^{+2}). However, the transition metals can form a variety of cations. They are said to have various oxidation states. When a metal loses an electron it is said to be oxidized. The oxidation state is the number of electrons it loses.

Consider iron. It can form the oxidation states $+2$ and $+3$, that is, it forms the cations Fe^{+2} and Fe^{+3}. Fe^{+2} is readily understandable. Fe can lose the two 4s electrons just like Ca to make the $+2$ oxidation state. In addition, Fe has six 3d electrons. Hund's Rule says that electrons will stay unpaired if possible. Five electrons can go into one of each of the five 3d orbitals. This half-filled configuration is particularly stable. Iron is one 3d electron past the half-filled 3d orbitals, so it will readily lose a 3d electron in addition to the two 4s electrons to give an oxidation state of $+3$. So Fe can form salts like $FeCl_2$ and $FeCl_3$.

In addition to giving rise to the first transition series (first group of transition metals), the 3d electrons are involved in another important molecular phenomenon. We discussed that oxygen will form two covalent bonds (share two electrons with other atoms) to obtain the Ne configuration. An example is water, H_2O. Sulfur, which is directly below oxygen, forms H_2S, analogous to H_2O. However, it can also form SF_6 through involvement of the 3d orbitals, which are close in energy to the 3p orbitals. There is no equivalent for oxygen because the first set of d orbitals, the 3d's, are much higher in energy than the 2s and 2p orbitals that are involved in bonding in the second row of the Periodic Table.

After the first series of transition metals are completed by filling the 3d orbitals, the next element is gallium (Ga). Ga is a metal, and like aluminum, it will form $+3$ ions. The configuration in which the 3d orbitals are completely filled is very stable, so Ga only forms $+3$ cations. The stability of the filled 3d orbitals can also be seen in zinc. Zn only forms $+2$ ions by losing the two 4s electrons. Following Ga are germanium (Ge), arsenic (As), and selenium (Se), which tend to form four, three, and two covalent bonds, respectively, to obtain the krypton (Kr) closed shell configuration. Like the elements immediately above Ge, As, and Se, additional bonds can be formed by involving the 4d electrons, which are very close in energy to the 4p orbitals. The next element is bromine, which is a halogen, and forms a -1 anion to obtain the Kr closed shell configuration. Finally, the row ends with krypton, which has a closed shell.

Larger Atoms and the Lanthanides and Actinides

The elements in the fifth row of the Periodic Table follow the same trends as those in the fourth row. The fifth row has the second series of transition metals. The elements in the sixth and seventh rows behave like those in the fourth and fifth rows except that they also have the lanthanides (first inner transition series) and actinides (second inner transition series). These come about by filling the 4f and 5f orbitals (see the many electron atom energy level diagram, Figure 11.1). The 4f (lanthanides) and 5f (actinides) orbitals (n = 4 and 5) are spatially much smaller than the 6s and 6p and 7s and 7p orbitals (n = 6 and 7) that are filled in the sixth and seventh rows because the principal quantum numbers, n, are smaller. The outermost electrons (largest principle quantum number) determine the chemical properties of atoms, that is, how many covalent bonds they make or what types of ions they form. Therefore, the 4f and 5f orbitals do not influence the chemical properties significantly. The lanthanides begin with lanthanum (La). The 4f energy levels are very close in energy to the 5d levels (see Figure 11.1). La comes after barium (Ba), which has two electrons in the 6s orbital. La has one more electron, which is actually in a 5d orbital. After La the 4f orbitals are filled. Lutetium (Lu, element 71) begins the third transition metal series. It has two electrons in the 6s orbital, 14 electrons in the 4f orbitals, and one electron in a 5d orbital. All of the lanthanides have chemical properties that are quite similar to La and Lu. In the same manner, the actinides begin with actinium (Ac). After filling the 5f orbitals with 14 electrons, lawrencium (Lr, the manmade element 103) begins the fifth transition metal series. All of the actinides have chemical properties that are very similar to Ac and Lr.

Most Elements Are Metals

The Periodic Table is color coded (shaded in Figure 11.4), with the elements divided into metals, semimetals (semiconductors), and nonmetals (insulators). (A detailed quantum theory explanation of why materials are metals, semiconductors, or insulators is presented in Chapter 19.) The Periodic Table shows that the vast majority of elements are metals. It is easy to see why this is. The left two columns are metals because they are comprised of elements that are either one or two s electrons past the previous noble gas closed shell configuration. They can readily give up these electrons to fall back to the closed shell configuration. Therefore, in solid form, it is easy to move electrons, and the solids are electrical conductors. Adding d electrons in the transitions series doesn't eliminate the ability of an element to give up its outermost (highest n) s electrons. The d electrons only add more electrons that can be lost under the right circumstances. Adding f electrons doesn't change things. Therefore, in addition to the two left-hand columns of elements, all of the transition series of elements are metals, usually called the transition metals. The inner transition series (addition of f electrons) are also metal. The elements that can lose three electrons to fall back to the previous closed shell configuration, like aluminum, are also metals. All of these together comprise most of the elements. The nonmetals are the group of elements in the upper right triangle-like block of elements in the Periodic Table. Some of these are the elements that tend to form covalent bonds by sharing electrons. They do not want to give up electrons. The halogens want to gain electrons or form covalent bonds. And the noble gases, by and large, do not want to gain or lose electrons or form covalent bonds. Therefore, all of these are nonmetals. If they are solids, the atoms do not want to give up electrons, a property necessary to conduct electricity. They are insulators. The small group of elements that form a diagonal block near the right side of the Periodic Table are semimetals or semiconductors. They are between the true metals and the nonmetals. Under some circumstances, they will conduct electricity. Silicon is the most well known and most technologically important of these semiconductors. Silicon is used in all of the microelectronics in our computers and other digital devices. In Chapter 19, we will discuss the differences among metals, insulators, and semiconductors, using the ideas of molecular orbital theory, which will be introduced in Chapter 12 and expanded on in the following chapters.

In this chapter, we used the many electron energy level diagram (Figure 11.1), and the rules for filling the atomic orbitals (the Pauli Principle, lowest energy first if possible, Hund's Rule) to discuss the Periodic Table. It was shown that very simple considerations can go a long way toward understanding the properties of the atomic elements and some aspects of chemical bond formation to make molecules. However, we did not use our ideas of quantum theory to discuss why chemical bonds should form and properties of molecules, such as their shapes, that arise from quantum considerations. We will begin the explication of the quantum theory of molecules with the simplest molecule, the hydrogen molecule H_2, in Chapter 12.

12

The Hydrogen Molecule
and the Covalent Bond

ONE OF THE GREAT TRIUMPHS of quantum mechanics is the theoretical explanation of the covalent bond. Two types of interactions hold atoms together, covalent bonds and ionic bonds. Ionic bonds are the type that occurs in a sodium chloride (NaCl) crystal. We know from Chapter 11 and our discussion of the Periodic Table that this salt crystal is composed of sodium cations, Na^{+1}, and chloride anions, Cl^{-1}. The ions in the crystal are held together by electrostatic interactions. Opposite charges attract. There are some complications because like charges repel, but it is possible to show that the attractions of the oppositely charged ions overcome the repulsions of the like charged ions. Such electrostatic interactions can be explained quite well with classical mechanics, although quantum theory is still needed to explain many properties in detail.

In contrast to ionic solids that are held together by electrostatic interactions, classical mechanics cannot explain the covalent bond. We saw in Chapter 11 that a hydrogen atom will tend to form one covalent bond with another atom to share one electron. This sharing brings the H atom to the helium closed shell configuration. But what is a covalent bond? Why do H atoms share electrons to form the H_2 molecule, but helium atoms do not share electrons to form the He_2 molecule? We will first investigate the nature of the covalent bond for the simplest molecule, H_2, and then expand the discussion of covalent bonding for more complicated molecules in subsequent chapters. By the end of this chapter it will be clear why H_2 exists and He_2 does not exist.

TWO HYDROGEN ATOMS THAT ARE FAR APART

Two hydrogen atoms, call them a and b, do not interact with each other when they are very far apart. When they are separated by a large distance, the electron on hydrogen atom a will only feel the Coulomb attraction to the proton in hydrogen atom a. The electron on hydrogen atom b will only interact with the proton on hydrogen atom b. We know how to describe these separated hydrogen atoms. Let's say both are in their lowest energy state, the 1s state. The electron is describe by the 1s wavefunction, which is an atomic orbital. It describes the probability amplitude of finding the electron in a given region of space. The square of the wavefunction gives the probability of finding the electron. The 1s state of the hydrogen atom was discussed in some detail in Chapter 10 (see Figures 10.2–10.4).

TWO HYDROGEN ATOMS BROUGHT CLOSE TOGETHER

Now consider what happens when we start bringing the two hydrogen atoms closer and closer together. When they get close together but not too close they start to feel each other. This distance will be made quantitative below. The electron on hydrogen atom a starts to have attraction for the proton of hydrogen atom b, and it has repulsion for the electron on hydrogen atom b. In the same manner, the electron on hydrogen atom b is attracted to the proton of hydrogen atom a and is repelled by the electron of hydrogen atom a. In addition, the positively charged protons of hydrogen atom a and b repel each other because like charges repel.

It is possible to solve the Schrödinger Equation for this problem. It can't be done exactly, but it can be done very accurately. What does the solution of the Schrödinger Equation give you? It gives the energies of the system, and it gives the wavefunctions. When we solved the particle in a box problem, we obtained the wavefunction for a single particle in a hypothetical one-dimensional box with infinite walls. When we solve the Schrödinger Equation for the hydrogen atom or other atoms, we get the energy levels and the atomic wavefunctions, the atomic orbitals. When we solve a molecular problem, we get the quantized energies of the molecular energy levels and the molecular wavefunctions. The molecular wavefunctions are usually called molecular orbitals. So for atoms we get atomic orbitals that describe the electron's probability distribution about the atomic nucleus. This is a probability amplitude wave. The molecular orbital describes the probability distribution for the electrons in the molecule relative to the atomic nuclei of the atoms that make up the molecule. For the hydrogen molecule there are two electrons and two atomic nuclei, the two protons.

THE BORN-OPPENHEIMER APPROXIMATION

A very useful way to understand bonding of hydrogen atoms as they come together to form the hydrogen molecule uses a concept called the Born-Oppenheimer Approximation. As discussed in Chapter 5, Born won the Nobel Prize in 1954 for the Born probability interpretation of the wavefunction. Oppenheimer made major contributions to the field of physics. He is probably best known as the physicist who led the Manhattan Project during World War II that developed and tested the first atomic bomb. The Born-Oppenhei-mer Approximation works by placing the two hydrogen atom nuclei (the two protons) a fixed distance apart. Start with a distance that you suspect to be so far apart that the hydrogen atoms don't feel each other. A quantum mechanical calculation of the energy is per-formed. If the atoms are far apart, then the energy will be just twice the energy of the 1s state of the hydrogen atom because you have two hydrogen atoms. Then you make the distance a little closer and do the calculation again. Then you make the distance even closer and do the calculation yet again. When the distance between the nuclei in the calculation becomes small enough, the atoms feel each other. If a chemical bond is going to form, that is, if the two hydro-gen atoms are going to combine into a hydrogen molecule, then the energy must decrease. To form a bond, the energy of the molecule must be lower than the energy of the atoms when they are separated far apart.

Figure 12.1 is a plot of the energy of two hydrogen atoms as they are brought closer and closer together. As mentioned, when the two H atoms are very far apart, they don't interact with each other. Each is just a hydrogen atom, and each has the energy of the hydrogen 1s atomic orbital. We are going to take this as the zero of energy. The hydrogen atom itself has a negative energy as described in Chapter 10. That energy represents the binding of the electron to the proton (nucleus). Here we are interested in the change in energy when the two H atoms interact. We want to understand the energy associated with chemical bond, so the zero of energy is the energy when there is no chemical bond. In Figure 12.1, the zero of energy is indicated in the diagram by the dashed line. This is the energy when the atoms are completely separated. The horizontal axis is the separation, r, of the two H atoms. As the H atoms are brought together the energy begins to decrease, then decreases more rapidly. The energy reaches a minimum at a separation, r_0 (see Figure 12.1). As the atoms are brought even closer together the energy increases

FIGURE 12.1. *A plot of the energy of two hydrogen atoms as they are brought close together. When the H atoms are very far apart, the energy of the system is the sum of the 1s orbital energies of two H atoms. This is taken to be the zero of energy, the dashed line. As the atoms come together, the energy decreases to a minimum. If they are brought even closer together, the energy increases rapidly.*

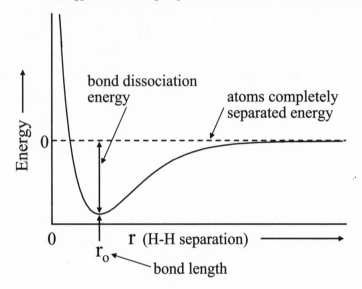

very rapidly, that is, when the atoms are too close, they repel each other. Because the energy goes down when the atoms are brought together, a chemical bond is formed between the two atoms.

The Bond Length Is the Distance That Gives the Lowest Energy

At the distance r_0, the energy is the minimum. r_0 is the separation of the H atoms that is the most stable (lowest energy). This distance is called the bond length. It is the separation between the two protons in a stable hydrogen molecule. The difference between the bottom of the "potential energy well" and the zero of energy is the dissociation energy. The dissociation energy is the amount of energy that would have to be put into a hydrogen molecule to break the chemical bond, which would produce two hydrogen atoms. The potential energy well for the hydrogen molecule is equivalent to a hole in the ground that a ball rolls into. The top of the hole is the zero of energy. The ball falls to the bottom of the hole to minimize the gravitational potential. Gravity pulls the ball down. To lift the ball out of the gravitational well requires energy, as the gravitational pull on the ball must be overcome. The deeper the hole, the more energy it will take to lift the ball out of the hole. With molecules, the deeper the potential energy well, the more energy it will take to get out of the well, that is, to break the chemical bond.

The distance scale along the r axis of Figure 12.1 is not shown. But it is interesting to discuss two distances. At what distance do the hydrogen atoms first really begin to feel each other? Figure 10.3 shows that the 1s orbital probability amplitude wave for a hydrogen atom becomes very small at a distance from the nucleus of approximately 3 Å (3×10^{-10} m). So one might expect that when two hydrogen atoms are a little closer than 6 Å they would start to interact. In Figure 12.1, the point where the potential energy curve (solid curve) just stops touching the zero of energy line (dashed line) is approximately 6 Å. So the atoms begin to feel each other when the atomic wavefunctions just begin to overlap significantly. The point r_0 is the location of the minimum of the potential energy curve. It is the length of the bond. Experiments and calculations have determined this distance to be 0.74 Å. If the atoms are further apart or closer together than this distance, the energy is higher. The potential energy curve shown in Figure 12.1 is from an actual quantum mechanical calculation. It is a relatively low-level calculation that can be done completely with pencil and paper; no computers are necessary. This approximate calculation gives $r_0 = 0.80$ Å, so it is a little off. If you want to see the monumental amount of math that goes into even this relatively simple calculation, see Michael D.

Fayer, *Elements of Quantum Mechanics,* Chapter 17 (New York: Oxford University Press, 2001). Much more complicated quantum theoretical calculations of the H_2 molecule can produce all of the properties of the hydrogen molecule with accuracy better than can be obtained through experimental measurements. Such accurate calculations are possible because the hydrogen molecule is so simple. For large molecules, experiments still beat calculations.

FORMING BONDING MOLECULAR ORBITALS

Figure 12.1 shows that a chemical bond will be formed between two hydrogen atoms to yield the H_2 molecule, but it doesn't show why. As mentioned in Chapter 11, a covalent bond involves the sharing of electrons by atoms. When a molecule is formed, the atomic orbitals combine to form molecular orbitals. For the hydrogen molecule, we start with two hydrogen atoms, H_a and H_b. Each has a single electron in an atomic 1s orbital. We will call these orbitals $1s_a$ and $1s_b$. These two atomic orbitals are represented in the top portion of Figure 12.2 as circles. This is a simple schematic of the delocalized electron probability amplitude wave shown in Figures 10.2 through 10.4. The lower portion of the figure shows what happens when the two atoms are brought together so they are separated by the bond length, r_0 (see Figure 12.1). Wavefunctions have signs. In this case, both signs are positive. The probability amplitude waves add together to form a molecular orbital. We discussed waves in some detail in Chapters 3 and 5. In Chapter 3, we saw that waves could be combined to have constructive or destructive interference. In Chapter 5, the interference of photons was explained in terms of the Born interpretation of the wavefunction as a probability amplitude wave. Here, the two atomic orbital electron probability amplitude waves combine constructively to form a molecular orbital. The molecular orbital is a probability amplitude wave. The absolute

FIGURE 12.2. *The upper portion is a schematic of two hydrogen atom 1s orbitals. These are actually delocalized electron probability amplitude waves, represented here simply as circles. The lower proportion shows what happens when the two atoms are brought together to form the H_2 molecule. The two atomic orbitals combine to make a molecular orbital.*

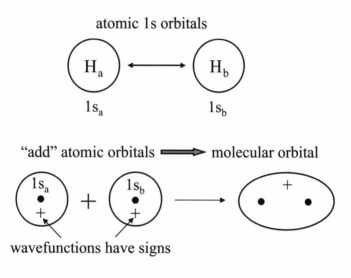

value squared of the wave gives the probability of finding the electrons in some region of space.

Figure 12.3 shows a one-dimensional plot of the probability distribution (square of the wavefunction) of the two atomic orbitals $1s_a$ and $1s_b$, and the square of the sum of the atomic orbitals. The atomic orbitals are centered on the H atom nuclei, which are separated by the bond length, r_0. The protons (nuclei) are positively charged and repel each other. However, the molecular orbital concentrates the negatively charged electron density between the nuclei and holds them together. The important feature of the molecular orbital is that the electrons no longer belong to one atom or the other. The molecular orbital describes a delocalized probability distribution for the two electrons. Both electrons are free to roam over

FIGURE 12.3. *A one-dimensional plot of the square of the two 1s orbitals that belonged to H atoms a and b (solid curves), and the square of the sum of the atomic orbitals, which is the square of the molecular orbital. The electron density is concentrated between the two nuclei.*

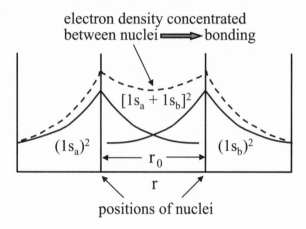

the entire molecule. The two electrons, which belonged to different atoms when the atoms were far apart, belong to the entire molecule. They are shared by the atoms, which are no longer independent.

BONDING AND ANTIBONDING
MOLECULAR ORBITALS

Something very important has been left out of the discussion so far. The Pauli Principle (Chapter 11) states that at most two electrons can be in any orbital. This is true of atomic orbitals or molecular orbitals. To form the hydrogen molecule, we began with two hydrogen atoms. Each hydrogen atom has a 1s orbital. According to the Pauli Principle, it would be possible to put four electrons in these two orbitals. The two hydrogen atoms only have two electrons, but it would not violate the Pauli Principle to add another electron to each of the atomic 1s orbitals. We added together the two 1s orbitals (constructive interference) to form one molecular orbital. The Pauli Principle tells us that we can put at most two electrons in this molecular orbital. We started with two atomic orbitals that could hold four electrons, but now we have one molecular orbital that can only hold two electrons. Something is missing. You never lose or gain orbitals or places for electrons when forming molecules. If you start with two atomic orbitals, then two molecular orbitals will be formed, which can hold four electrons.

In Figures 12.2 and 12.3, the two 1s hydrogen atomic orbitals were added with the same sign to produce a molecular orbital that concentrates the electron density between the two atomic nuclei. The 1s orbitals are probability amplitude waves and can also be added with opposite sign. When the atomic orbitals are added with opposite sign, they destructively interfere. The addition with opposite sign is shown in Figure 12.4. Because the signs of the two atomic orbitals are opposite, there must be a place where the positive wave exactly cancels the negative wave. This is a node, as we discussed before for the particle in the box wavefunctions and atomic orbitals. As can be seen in the schematic in Figure 12.4, the destructive interference between the atomic orbitals and the result-

FIGURE 12.4. *The left side is a schematic of two hydrogen atom 1s orbitals that are being added together. Note that the probability amplitude waves have opposite signs. The two atomic orbitals combine to make a molecular orbital. Because of the opposite signs, there is destructive interference, in contrast to Figure 12.2.*

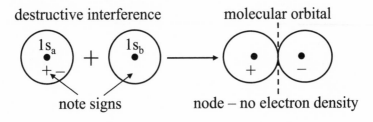

ing node pushes the electron density out from between the atomic nuclei. The negatively charged electrons are no longer screening the positively charged nuclei, which repel. In contrast to the curve shown in Figure 12.1, as the two hydrogen atoms are brought together from far apart, the energy goes up rather than down.

The molecular orbital that arises from constructive interference of the 1s atomic orbitals (see Figures 12.2 and 12.3), which is responsible for the H_2 chemical bond, is called a bonding molecular orbital, or bonding MO. The molecular orbital that arises from destructive interference between the atomic orbitals is called an antibonding MO because it does not give rise to a bond, and in fact increases the energy as the atoms are brought together.

Figure 12.5 is a schematic of the energy curves for the bonding and antibonding MOs of the H_2 molecule. As discussed in connection with Figure 12.1, as the two hydrogen atoms are brought together, the energy goes down and reaches a minimum before increasing again. This is the bonding MO curve. In contrast, the curve for the antibonding MO shows that the energy increases as the two atoms get close enough together to feel each other. The energy continues to increase as the atoms get progressively closer together. There is no reduction in energy. If the electrons are in the antibonding MO, the two atoms will not form a bond because the energy of the system is always higher than having separated atoms.

PUTTING ELECTRONS IN
MOLECULAR ORBITALS

We started with two atomic orbitals, $1s_a$ and $1s_b$, that were associated with two hydrogen atoms, H_a and H_b. These two atomic orbitals give rise to two molecular orbitals, the bonding and antibonding MOs. The rules we used for filling the atomic orbitals with electrons apply here as well. The Pauli Principle says no more than two electrons can go into an MO and they must have opposite spins (spins

FIGURE 12.5. *A schematic plot of the energy curves of the bonding and antibonding molecular orbitals for two hydrogen atoms as they are brought closer together. In contrast to the bonding MO, the energy of the antibonding MO increases as the atomic separation is decreased.*

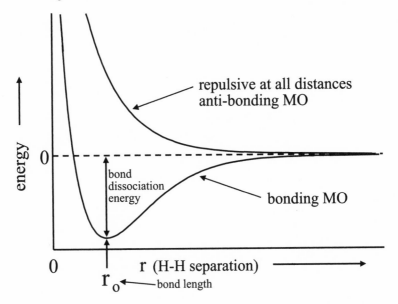

paired, one up arrow and one down arrow). Electrons go into the lowest energy levels first so long as this doesn't violate the Pauli Principle. Hund's Rule states that electrons will be unpaired if that doesn't violate the first two rules. We will not need Hund's Rule until the next chapter when we talk about larger molecules. Now we are in a position to see why the hydrogen molecule H_2 exists, but the helium molecule He_2 does not exist.

THE HYDROGEN MOLECULE EXISTS BUT THE HELIUM MOLECULE DOESN'T

At the atomic separation that corresponds to the bond length, that is, the separation you find in the actual molecule, the bonding MO is always lower in energy than the separated atoms and the anti-bonding MO is always higher in energy. This is a rigorous result from quantum mechanics. It is a good approximation to say that the energy decrease of the bonding MO is equal to the energy increase of the antibonding MO.

A simple diagram that is used to reflect the atomic orbitals coming together to form molecular orbitals is shown in Figure 12.6. We will use this type of diagram in subsequent chapters. The two 1s atomic orbitals, one for each H atom, are depicted on the left and right sides of the figure. The lines through them are the zero of energy for the molecular orbitals. That is, the lines are the energy of the atoms when they are so far apart that they do not feel each other. In the center are the energy levels of the bonding and antibonding MOs. These are called σ^b (b for bonding) and σ^* (* for

FIGURE 12.6. *An energy level diagram that represents the combination of the two atomic 1s orbitals to form the bonding and antibonding MOs at the bond length separation r_0, which is the distance of the energy minimum of the bonding MO. The bonding MO is lower in energy and the antibonding MO is higher in energy than the atomic orbitals by the same amount. The bonding MO is called σ^b, and the antibonding MO is called σ^*.*

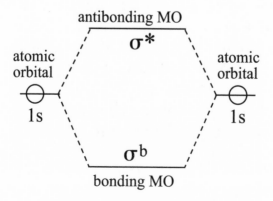

antibonding), and σ is the Greek letter sigma. σ designates a certain type of bond, called a σ bond, that will be discussed in Chapter 13. The dashed lines connecting the atomic orbitals to the MOs are used to indicate that both atomic orbitals combine to make both MOs when the atoms form the molecule.

Figure 12.6 shows the MO energy level diagram for the two energy states that are involved in forming a hydrogen molecule. We have not "put in" the two electrons yet. This diagram is analogous to the many electron energy level diagram, Figure 11.1. We have the energy levels, but now we need to put in the electrons to see what happens. There are two electrons, one from each of the hydrogen atoms. We know that electrons are placed in the lowest possible energy level so long as the number of electrons does not violate the Pauli Principle, that is, at most two electrons can be in any given orbital with spins paired. This applies to MOs as well as atomic orbitals. Figure 12.7 shows the MO energy level diagram with the two electrons (arrows). The two electrons go into the lowest energy level, σ^b, with spins paired. When the atoms are well separated, the electrons have the energy represented by the lines though the 1s atomic orbitals. The bonding MO has substantially lower energy. It is this reduction in energy that holds the molecule together. The two electrons are in a molecular orbital. Neither is associated with a particular atom. A covalent bond involves sharing of the electrons by the atoms.

Why doesn't the helium molecule, He_2, exist? Two separated He atoms each have 1s orbitals with two electrons in them. Therefore the MO diagram is the same as that shown in Figure 12.6. However, now we need to put four electrons into the MO energy levels. Figure 12.8 shows the MO diagram with the four electrons. The first electron goes into the bonding MO because it is the lowest energy state. The second electron also goes into the bonding MO with the opposite spin of the first. The Pauli Principle says that no two electrons can have all of their quantum numbers the same. The two electrons

FIGURE 12.7. *The MO diagram for the hydrogen molecule. The two electrons (arrows), one from each hydrogen atom, go into the lowest energy level with their spins paired. The energy is lower than the separated atoms. An electron pair sharing bond is formed.*

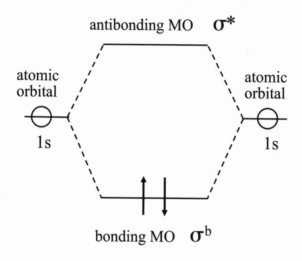

in the bonding MO have different spin quantum numbers, $s = +1/2$ and $s = -1/2$. There are only these two spin quantum numbers, so the third electron cannot go into the bonding MO. It must go into the next lowest energy level, which is the antibonding MO. The fourth electron can also go into the antibonding MO with the opposite spin. There are two electrons in the bonding MO and two electrons in the antibonding MO. The two electrons in the bonding MO lower the energy relative to the separated atoms, but the two electrons in the antibonding MO raise the energy just as much as the bonding electrons lowered the energy. The net result is that there is no decease in energy relative to the separated atoms. A molecule is held together because the bonded atoms have lower energy than the separated atoms. For helium atoms, there is no reduction in energy that would produce a stable configuration.

FIGURE 12.8. *The MO diagram for the hypothetical helium molecule. There are four electrons (arrows), two from each helium atom. Two go into the bonding MO. Because of the Pauli Principle, the other two go into the antibonding MO. There is no net reduction in energy and, therefore, no bond.*

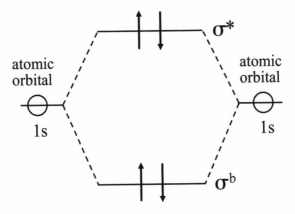

Therefore, there is no bond. We will see this same type of behavior for the noble gas neon in the next chapter.

To see the predictive capabilities of the simple diagrams like those shown in Figures 12.7 and 12.8, consider four molecules, or possible molecules. They are the hydrogen molecule ion H_2^+, the hydrogen molecule H_2, the helium molecule ion He_2^+, and the helium molecule He_2. H_2^+ is composed of two hydrogen nuclei (protons) and one electron. Like the atomic cation ion Na^+, it is positively charged because it has one fewer electrons than it has protons. He_2^+ is the molecular ion composed of two helium nuclei with two protons in each nucleus and three electrons. So it has four positive charges (four protons) and three negatively charged electrons.

Figure 12.9 shows the MO energy level diagrams for the four molecules. The atomic energy levels have been omitted. H_2^+ has only one electron, so it goes into the lowest energy level, the bond-

FIGURE 12.9. *The MO energy level diagrams for four molecules, the hydrogen molecule ion H_2^+, the hydrogen molecule H_2, the helium molecule ion He_2^+, and the helium molecule He_2.*

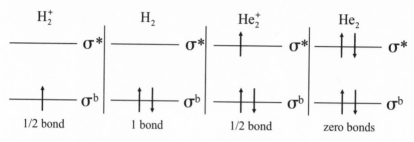

ing MO. The energy is lower than the separated atoms, but by only approximately half as much as for the H_2 molecule, which has two electrons in the bonding MO. H_2 has a full covalent bond. We say it has a bond order of 1. He_2^+ has a bond order of 1/2. He_2^+ has three electrons. The first two can go into the bonding MO, but, because of the Pauli Principle, the third electron must go into the antibonding MO. Two electrons lower the energy relative to the separated atoms, but one electron raises the energy. There is a net lowering of the energy. The He_2^+ molecular ion exists in nature, and it has a bond order of 1/2. As discussed, He_2 has two bonding electrons and two antibonding electrons. It has no bond. The bond order is zero. The He_2 molecule does not exist.

Table 12.1 gives some quantitative information about these four molecules. It gives the number of bonding electrons, the number of antibonding electrons, and the net, which is the number of bonding electrons minus the number of antibonding electrons. It also gives the bond order. The last two columns are of particular interest.

These are the results of experimental measurements on the molecules. First consider the bond length. The lengths are in Å (Ångstroms, 10^{-10} m). The H_2^+ molecule has a bond order of 1/2 and a bond length of 1.06 Å. In contrast, H_2, which has a full bond

TABLE 12.1. *Properties of the hydrogen molecule ion* H_2^+, *the hydrogen molecule* H_2, *the helium molecule ion* He_2^+, *and the helium molecule* He_2.

	Bonding electrons	Antibonding electrons	Net	Bond order	Bond length	Bond energy
H_2^+	1	0	1	$1/2$	1.06	4.2
H_2	2	0	2	1	0.74	7.2
He_2^+	2	1	1	$1/2$	1.08	5.4
He_2	2	2	0	0	none	none
					(Å)	$(10^{-19} J)$

with bond order 1, has a bond length of 0.74 Å. The additional bonding electron in the H_2 bonding MO holds the atoms together tighter and therefore, closer. He_2^+ has a bond order of 1/2 and a bond length of 1.08 Å, which is only a little longer than H_2^+. Of course, He_2 is not a molecule, so it does not have a bond length. The last column is the bond energy in units of 10^{-19} J. The relative strength of the bonds is interesting. The H_2 molecule with a bond order of 1 has a considerably stronger bond than the two molecular ions, which have bond orders of 1/2. These simple MO diagrams allow us to see if a bond will exist, and they give information on how strong the bond will be.

In this chapter, we have used the ideas of molecular orbitals to look at the simplest molecules. The discussion only involved atoms that have 1s electrons. All other atoms and molecules involve more electrons and more orbitals. In the next chapter, the ideas introduced here will be used to examine diatomic molecules involving larger atoms, such as the oxygen molecule, O_2, and the nitrogen molecule, N_2. These two molecules are the major components of the air we breathe.

13

What Holds
Atoms Together:
Diatomic Molecules

THE HYDROGEN MOLECULE is a diatomic molecule, that is, a molecule composed of only two atoms. Our investigation of hydrogen revealed how atoms can combine their atomic orbitals to form molecular orbitals. We need to extend our discussion of molecular orbitals to learn how more complicated molecules are built up from atoms. First, we will consider other diatomic molecules, for example, N_2, O_2, F_2, and HF. N_2, O_2, and F_2 (nitrogen, oxygen, and fluorine) are called homonuclear diatomics because the two atoms are the same. HF (hydrogen fluoride) is a heteronuclear diatomic because the two atoms are different. Examination of homonuclear diatomics will let us go beyond the hydrogen molecule, which is a special case. The nature of molecular orbitals for heteronuclear diatomics is an important step toward the study of polyatomic molecules that compose most of the molecular substances that surround us, from alcohol to fats.

The hydrogen molecule is the only neutral molecule that only uses electrons in its 1s orbitals to form a chemical bond. The electrons that an atom employs in bonding are called the valence electrons. In the molecules N_2, O_2, F_2, and HF, 2s and 2p orbitals are involved in bonding. The 2s and 2p electrons are the valence electrons. N, O, and F are in the second row of the Periodic Table. Atoms in the third row of the Periodic Table, such as P, S, and Cl (phosphorus, sulfur, and chlorine), will have 3s and 3p valence electrons involved in bonding. Atoms in the third and higher rows of the Periodic Table can also employ d electrons in forming chemical bonds. Here we will focus on the second row elements. The second row elements are of great importance, and the ideas that will be developed are sufficiently general to cover the nature of chemical bonding of heavier elements.

SIGMA (σ) AND PI (π) BONDS

As shown in Figure 12.2, when two hydrogen atoms form the H_2 molecule, the two hydrogen 1s orbitals combine to form the bonding molecular orbital. There is electron density along the line that connects the nuclei. A σ (sigma) bonding or antibonding molecular orbital has electron density along the line connecting the nuclei. In H_2, we say that a σ bond is formed using a σ bonding MO. s orbitals always form σ bonds. There is no way to bring together two s orbitals and not have electron density along the line connecting the nuclei. This is not true for p orbitals.

Because of their shapes, there are two ways for a pair of p orbitals to come together, as is illustrated in Figure 13.1. The drawings in Figure 13.1 show the p orbitals very schematically. These are actually probability amplitude waves that have a diffuse probability distribution of finding the electron some distance from the nucleus. Here the outline gives a representation of the general shape of a p orbital. Better illustrations are shown in Figure 10.7. Recall that a p orbital has a nodal plane between the two lobes. A nodal plane is a plane in which the probability of finding the electron is zero. For a p_z orbital, the nodal plane is the xy plane (see Figure 10.7). The probability of finding an electron in some region of space is fre-

FIGURE 13.1. *A pair of p orbitals brought close to each other. Upper portion: the orbitals are brought together end to end. There is electron density along the line connecting the nuclei. Bottom portion: the orbitals are brought together side to side. There is no electron density along the line connecting the nuclei.*

Lobes of p orbitals point at each other.

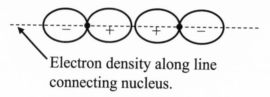

Electron density along line
connecting nucleus.

Lobes of p orbitals are side to side.

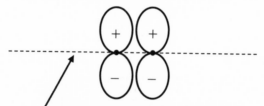

No electron density along line
connecting nuclei.
Nodal plane contains nuclei.

quently called the electron density. A high density means there is a high probability of finding the electron.

The upper portion of Figure 13.1 shows two p orbitals brought close to each other end to end. The lobes are pointing at each other. The nuclei are represented by the dots. The dashed line is the line connecting the nuclei. There is clearly electron density along the line connecting the two nuclei. The lower portion of the figure shows the two p orbitals brought close to each other side to side. The nodal plane is perpendicular to the plane of the page. The nuclei lie in the nodal plane. There is no electron density along the line connecting the nuclei. The lobes of the p orbitals have signs. One lobe is positive and the other negative. In Figure 13.1, the positive lobes were brought together next to each other in both the top and bottom portions of the figure.

SIGMA MOLECULAR ORBITALS

When the orbitals are brought sufficiently close together, they can form bonding and antibonding molecular orbits. First we will look at molecular orbitals formed from s and p atomic orbitals that give σ bonding and antibonding MOs. A σ MO has electron density along the line joining the nuclei. As discussed above, s orbitals can only form σ orbitals because of their spherical shape. p orbitals can also form σ MOs. Figure 13.2 shows σ bonding and antibonding MOs formed from both s and p orbitals. The upper part of the figure depicts the two possible ways that a pair of s orbitals can be combined. The s orbitals are waves and can have either a + or − sign associated with them. In the top portion, both s orbitals have + signs. When they combine, the s orbital waves constructively interfere to produce a σ bonding MO. In the second line of the figure, one s orbital is + and the other is −. When they combine, they destructively interfere to form an antibonding MO. The bonding MO concentrates the electron density between the nuclei, while the antibonding MO pushes the electron density to the outside, reducing the negative electron density between the nuclei. The positively charged nuclei repel each other more strongly, making this configuration antibonding.

The bottom portion of Figure 13.2 shows the results of combining two p orbitals to form σ molecular orbitals. The σ p bonding MO is generated by overlapping the + lobe of one p orbital with the + lobe of the other p orbital. There is constructive interference between the + lobes, resulting in high electron density between the

FIGURE 13.2. *Upper portion: a pair of s orbitals are overlapped in two ways to give σ bonding (constructive interference) and σ antibonding (destructive interference) molecular orbitals. Lower portion: a pair of p orbitals are overlapped in two ways to give σ bonding (constructive interference) and σ antibonding (destructive interference) molecular orbitals. In all cases, there is electron density along the line through the nuclei.*

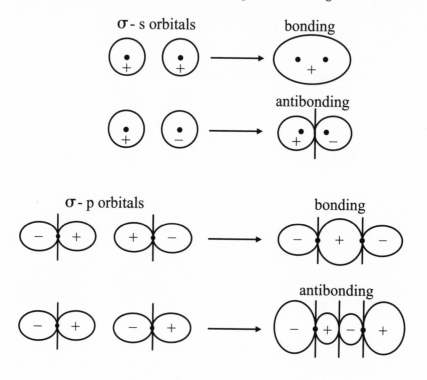

atomic nuclei. There are two nodal planes that are perpendicular to the page. These are the two nodal planes that came from the two atomic p orbitals. In contrast, in the bottom line of the figure, the positive lobe of one p orbital is overlapped with the negative lobe of the other p orbital. The result is destructive interference, producing the antibonding MO. The electron density is pushed to the outside, and it is greatly reduced between the two nuclei. In addition to the two nodal planes that come from the atomic orbitals, there is a third

nodal plane that arises because there is complete destructive interference between the positive and negative lobes of the two atomic p
orbitals. In these bonding and antibonding MOs formed from
atomic p orbitals, there is electron density along the line connecting
the nuclei. Therefore, these are σ MOs.

PI MOLECULAR ORBITALS

s orbitals can only form σ MOs, but p orbitals can form σ MOs and
another type of molecular orbital called π (the Greek letter pi) MOs.
When p atomic orbitals come together end to end, they form σ
MOs. When they come together side to side, they form π MOs, as
shown in Figure 13.3. The upper portion of the figure shows two p
orbitals forming a bonding molecular orbital. The positive lobe of
one atomic orbital overlaps the positive lobe of the other, and the
negative lobes overlap. There is constructive interference for both
the positive and negative lobes. As can be seen in the figure, there
is a great deal of electron density in the area between the two nuclei.
However, there is no electron density directly along the line connecting the nuclei. There is a nodal plane that is perpendicular to
the plane of the page because each atomic orbital has a nodal plane
perpendicular to the page. The nodal plane contains the nuclei. In
spite of the nodal plane, all of that electron density immediately
above and below the line connecting the nuclei reduces the repulsion of the positive nuclear changes. The energy is lower than the
separated atoms resulting in a π bonding MO.

The lower portion of Figure 13.3 shows the π antibonding MO.
The two p atomic orbitals come together side to side, but the positive lobe of one orbital overlaps the negative lobe of the other orbital,
and vice versa. The result is destructive interference between the
lobes giving rise to the π antibonding MO. The antibonding MO
has much less electron density between the nuclei. The result is
that the energy is higher than the separated atoms, so this configuration of atomic orbitals produces an antibonding MO.

FIGURE 13.3. *Upper potion: a pair of p orbitals are overlapped side to side to give a π bonding molecular orbital (constructive interference). There is no electron density along the line connecting the nuclei. Lower portion: a pair of p orbitals are overlapped side to side to give a π antibonding molecular orbital (destructive interference). Note the signs of the lobes of the p atomic orbitals. The antibonding MO has a node between the nuclei.*

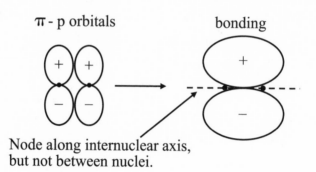

π - p orbitals   bonding

Node along internuclear axis, but not between nuclei.

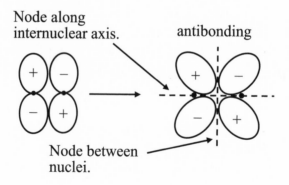

Node along internuclear axis.   antibonding

Node between nuclei.

BONDING IN DIATOMIC MOLECULES:
THE FLUORINE MOLECULE

We are now ready to discuss bonding for diatomics with atoms other than hydrogen. Let's start with the diatomic, F_2, the fluorine molecule. We will use the same approach as we did for H_2, but now there are more orbitals and more electrons involved. We imagine bringing two F atoms together and stopping at the point where the energy is lowest. This is the separation of the two F atoms when they are bonded (assuming they form a bond), as in Figure 12.5. We can draw an energy level diagram, as in Figure 12.6. We need to define the axis along which the two atoms approach each other because we have p_z, p_x, and p_y orbitals. It matters whether we bring p orbitals together end to end or side to side. When the two atoms (labeled a and b) approach along the z axis (see Figure 13.4), the p_z orbitals will come together end to end, while the p_x and p_y orbitals will come together side to side. Therefore, the p_z atomic orbitals will form σ MOs and the p_x and p_y atomic orbitals will form π MOs.

Figure 13.5 shows the energy level diagram for two F atoms brought together along the z axis. In the diagram, the energy levels for the atomic orbitals for the two atoms (a and b) are shown on the right and left sides of the diagram. The corresponding bonding (b) and antibonding (*) MOs are shown in the center. σ MOs formed from s atomic orbitals have a subscript s. σ MOs formed from p_z atomic orbitals have a subscript z, and π MOs formed from p_x or p_y atomic orbitals have a subscript x or y. The bonding MO is always lower in energy than the atomic orbitals that formed it, and the antibonding MO is always higher in energy. The three p atomic orbitals have the same energy. When quantum states have the same energy, they are said to be degenerate. In the diagram, the three p atomic orbitals are shown as three closely spaced lines even though they are degenerate. As shown, only the matching atomic orbitals

FIGURE 13.4. *Two atoms are brought together along the z axis. P_z orbitals will approach end to end; p_x and p_y orbitals will approach side to side.*

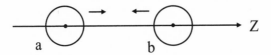

FIGURE 13.5. *The energy level diagram for two fluorine atoms, a and b, brought together to form molecular orbitals. The atomic orbital energies are on the right and left sides. The bonding (b) and antibonding (*) MO energy levels are in the center. There are σ and π MOs. The three atomic p orbitals have the same energies. These are shown as three closely spaced lines. The spacings between the levels are not to scale.*

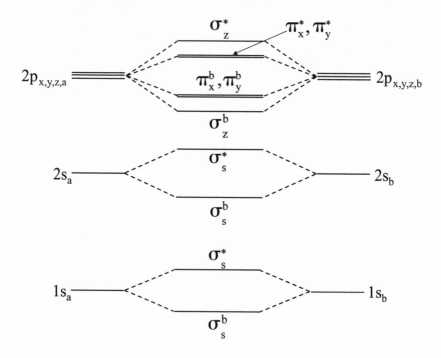

of the same energy combine to make MOs. Quantum theory gives this result. States of identical energy can readily combine to make superposition states. In this case, atomic orbitals of identical energy on two different atoms can combine to make molecular orbitals. In general, only atomic states with similar energy can combine to make MOs. This will be important when we discuss heteronuclear diatomics below. For the homonuclear diatomics, the atomic orbitals are identical in energy. In the diagram, the three p orbitals on each atom, a total of six atomic orbitals, combine to form six molecular orbitals. The p_z atomic orbitals produce the σ bonding and antibonding MOs, which are different in energy from the bonding and antibonding π_x and π_y MOs formed from the p_x and p_y atomic orbitals. However, the π_x and π_y bonding MOs have the same energy, and the π_x and π_y antibonding MOs have the same energy. The degenerate pairs of π MOs are shown as two closely spaced lines.

Fluorine has nine electrons. So a fluorine atom will have two electrons in the 1s orbital, two electrons in the 2s orbital, and five electrons in the three 2p orbitals. Two F atoms have a total of 18 electrons. We now need to place the 18 electrons in the proper molecular orbitals in a manner equivalent to the way we placed the electrons in the atomic orbitals when we were building up the Periodic Table in Chapter 11 and the hydrogen molecule in Chapter 12. As before, we need to follow the three rules for filling the MOs. First is the Pauli Principle, which states that at most two electrons can be in an orbital and the two must have opposite spins. Opposite spins are represented by one up arrow and one down arrow. Second, electrons are placed in the lowest energy level first, consistent with the Pauli Principle. Third is Hund's Rule, which states that electrons will not pair their spins if possible. In F_2, Hund's Rule will not change the results of placing the electrons in the proper orbitals. When we consider oxygen, O_2, it will be important.

Figure 13.6 is the MO energy level diagram for F_2 with the electrons placed in the proper orbitals. The atomic orbital energy levels shown in Figure 13.5 are not included. Only the MO energy levels are shown. The first two electrons go into the σ bonding MO formed from 1s orbitals. The next two electrons go in the σ antibonding MO formed from 1s orbitals. The electrons in the bonding MO are lower in energy than the atomic orbitals of the separated atoms, but the electrons in the antibonding MO are the same amount higher in energy. Therefore, these four electrons do not contribute to bonding

FIGURE 13.6. *The molecular orbital energy level diagram for the F_2, fluorine diatomic molecule. The atomic orbital energies are not shown. Two fluorine atoms have 18 electrons. These have been placed in the orbitals following the rules discussed for atomic orbitals in Chapter 11. There is one more filled bonding MO than antibonding MO. F_2 has a single bond.*

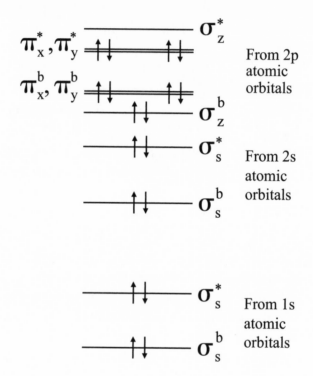

in F_2. The next four electrons go in the σ bonding and antibonding MOs formed from the 2s atomic orbitals. Again, these do not contribute to bonding because there are two electrons in the bonding MO and two electrons in the antibonding MO.

Now the p electrons come into play. There are a total of 10, five from each F atom. The first two go into the σ bonding MO formed from the p_z atomic orbitals. Then four electrons go into the π_x and

π_y bonding MOs. Four electrons can go into these π molecular orbitals because there are two MOs, and each can take two electrons according to the Pauli Principle. The last four electrons go into the π_x and π_y antibonding MOs. The four electrons in the π antibonding MOs cancel the bonding effect of the four electrons in the π bonding MOs. Therefore, the π electrons do not contribute to bonding. However, nothing cancels the two electrons in the σ bonding MO formed from the p_z atomic orbitals because there are no electrons in the corresponding antibonding MO. The net result is that one pair of bonding electrons is not canceled out, so F_2 has a bond order of 1 like H_2. We say that F_2 has a single bond, and it is a σ bond. The single covalent bond comes from having two electrons in a bonding MO. Molecular orbitals are probability amplitude waves that span the entire molecule. The atomic nuclei share these electrons.

THE NEON MOLECULE DOESN'T EXIST

We can use the same MO energy level diagram shown in Figure 13.6 to consider the hypothetical diatomic neon molecule, Ne_2. Neon is immediately to the right of fluorine on the Periodic Table. Figure 13.7 shows the results of placing the 20 electrons from two neon atoms into the MO energy level diagram. The first 18 electrons go in the same as in F_2. However, there are two more electrons, and they must go in the σ_z^* antibonding MO. Therefore, for every pair of electrons in a bonding MO there is a pair of electrons in an antibonding MO. The result is no bonding. The molecule Ne_2 does not exist. Other noble gas homonuclear diatomics do not exist. The example of Ne_2 indicates why. A noble gas atom has a closed shell. Two noble gas atoms have just the right number of electrons to fill all of the bonding and antibonding MOs. Therefore, no net bonds are formed.

FIGURE **13.7.** *The MO energy level diagram for the hypothetical mole-cule, Ne_2. Two neon atoms have 20 electrons. There are the same number of bonding and antibonding electrons, so there is no bond. Ne_2 does not exist.*

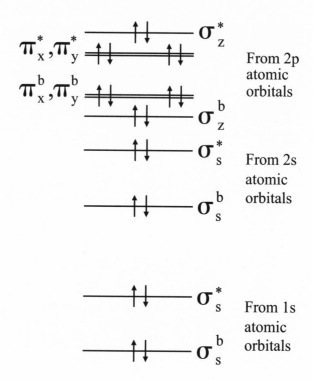

THE OXYGEN MOLECULE: HUND'S RULE MATTERS

One atom to the left of fluorine on the Periodic Table is oxygen. O_2 is an important example that introduces a couple of new ideas. Fig-ure 13.8 shows the MO energy level diagram filled with O_2's 16 electrons, eight from each oxygen atom. The bonding and antibond-ing MOs arising from the 1s and 2s orbitals are filled. These do not contribute to bonding. There are two electrons in the σ_z^b bonding MO and none in the corresponding antibonding MO. In addition,

FIGURE 13.8. *The MO energy level diagram for oxygen, O_2. There is one extra pair of σ bonding electrons and one extra pair of π bonding electrons. O_2 has a double bond. Note the unpaired electrons in the π antibonding MOs.*

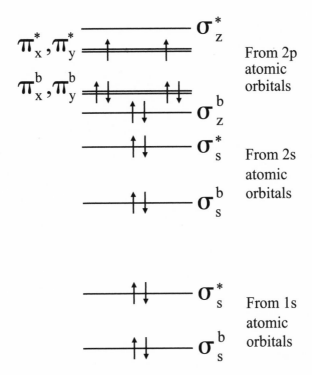

there are four electrons in the two π bonding MOs but only two electrons in the π antibonding MOs. The result is one σ bond and one π bond. Oxygen has a bond order of 2. It has a double bond. As will be discussed further below, a double bond is stronger and shorter than a single bond.

O_2 is the first example where Hund's Rule comes into play and is important. Note that in filling the energy levels with electrons, the last two electrons have their spins unpaired. It is possible to have unpaired spins without violating the Pauli Principle because there are two distinct π antibonding MOs. π_x^* comes from the side-to-side overlap of the two p_x atomic orbitals (see Figure 13.3), and π_y^*, comes from the side-to-side overlap of the two p_y atomic orbitals. Hund's Rule says that the electrons will go into orbitals unpaired if the result doesn't violate the Pauli Principle and doesn't require using a much higher energy orbital. The two π antibonding MOs have identical energy, so Hund's Rule comes into play.

An electron has a magnetic moment. In some sense it acts like a tiny bar magnet. It has a north pole and a south pole. The term spin for the electron quantum number comes from classical mechanics. In classical mechanics, a charge distribution that is spinning has a magnetic moment. An electron is a probability amplitude wave. It has a delocalized charge distribution. It has a magnetic moment, but it is not literally spinning. That is a classical idea. Dirac, who gave us the concept of absolute size (see Chapter 2), also showed why an electron has a magnetic moment by combining quantum theory and Einstein's Theory of Relativity. Although an electron is not actually spinning, the name stuck, and the magnetic moment of the electron is important.

When two electron spins are paired, the north pole of one little magnet is matched with the south pole of the other. The magnetic property of one electron cancels the magnetic property of the other. However, in O_2, two of the electrons are not paired. Their spins point in the same direction. The result is that the O_2 molecule is referred to as being paramagnetic. It will respond to a magnet. O_2 is a gas at room temperature, but if you make it very cold, below $-183°$ C ($-297°$ F) it will become a liquid. Water above 100° C is a gas, but if you cool it below 100° C, it becomes a liquid. O_2 is the same, but you need to make it much colder. It is possible to put liquid O_2 in a test tube hanging from a string. If you bring a magnet up to the test tube, you can actually pull the test tube around with the magnet. The electron spins (little bar magnets) of O_2 are lined up by the magnetic field of the external macroscopic magnet. These lined-up little bar magnets add together to make the liquid O_2 magnetic, and the test tube of O_2 is attracted to the external magnet. The correct prediction that O_2 is paramagnetic from the analysis we have done using the MO energy level diagram is remarkable. The magnetic moment of O_2 is strictly a quantum effect, and our prediction that O_2 is paramagnetic comes from the application of Hund's Rule. By following some rules, we drew some lines to represent the energy levels. Then following more rules, we put some arrows up and down on the energy level lines (we put in the electrons). Using these lines and arrows we can say that the oxygen molecule is magnetic while the fluorine and nitrogen molecules are not.

THE NITROGEN MOLECULE

Figure 13.9 shows the filled MO energy level diagram for nitrogen, N_2. The nitrogen atom is immediately to the left of oxygen on the Periodic Table. Note that there is a switch in the ordering of the bonding MOs derived from the p electrons. Detailed quantum mechanical calculations give the ordering and the actual energies of the MO energy levels. For nitrogen, the ordering is different than for O_2 and F_2. The nitrogen atom has seven electrons, so N_2 has a total of 14 electrons. As with F_2 and O_2, the 1s and 2s electrons do not play a role in bonding because they fill both the bonding and antibonding MOs. Filling these MOs uses up eight of the 14 electrons. The other six electrons fill the three bonding MOs, one σ MO and two π MOs. There are no electrons in the π antibonding MOs or the σ antibonding MO formed from the p_z orbitals. Therefore, N_2 has a bond order of 3, a triple bond. A triple bond will be stronger and shorter than a double or single bond. Note that there are no unpaired electrons in N_2. N_2 is not paramagnetic. At low temperature, below $-196°$ C ($-320°$ F), N_2 is a liquid. However, you cannot move a test tube of liquid N_2 with a magnet because it does not have unpaired spins.

SINGLE, DOUBLE, AND TRIPLE BONDS

In discussing bonding in Chapter 11 based on an atom's position in the Periodic Table, we used the idea that an atom would form covalent bonds to share electrons so that it could reach the noble gas configuration. For the second row elements we have been discussing here, nitrogen, oxygen, and fluorine, the noble gas is neon. We said that F, which is one electron from the Ne configuration, would share one electron with another atom. O, which is two electrons from the Ne configuration, would share two electrons, and N, which is three electrons from Ne, would share three electrons. Here we saw that F_2 has a single bond, O_2 has a double bond, and N_2 has a triple bond. The type of bond between atoms, that is single, double, or triple, can be indicated as $F-F$, $O=O$, and $N\equiv N$. It is through the bonds that atoms share electrons. A covalent bond is an electron pair sharing bond. A double bond shares two pairs of electrons. A triple bond shares three pairs of electrons. When a bonding MO is exactly canceled by its corresponding antibonding MO, the electrons are not really shared by the atoms. They are in molecular orbitals, but the bonding MO produces constructive interference of the probability amplitude waves and the antibonding MO generates destructive interference. These cancel each other. The electrons in this situation are referred to as lone pairs. These are pairs of electrons that do not contribute to bonding. It is the F_2 single bond, a shared pair of electrons, that provides the extra electron each F atom needs to give it the Ne configuration. In O_2, the double bond (sharing of two pairs of electrons) provides two extra electrons for each O atom to give them the Ne configuration. In N_2, the triple bond (sharing of three pairs of electrons) provides three extra electrons for each nitrogen atom to give them the Ne configuration.

In the series F_2, O_2, and N_2, we have a single bond, double bond, and triple bond. The electron sharing gives each atom the Ne configuration. The next element to the left of nitrogen in the Periodic Table is carbon. One might assume that carbon would make a quadruple bond to form C_2 and get to the Ne configurations. C_2 doesn't exist as a stable molecule. You can see why by looking at Figure 13.9, the MO diagram for N_2, and removing the two electrons with the highest energy, the σ_z^b bonding MO. This would be the C_2 electron configuration. However, it would have a double bond from the four electrons in the two π bonding MOs, not a quadruple bond. These two bonds would mean that the carbons in C_2 would only have gained two extra electrons through sharing, not the four extra

FIGURE 13.9. *The MO energy level diagram for nitrogen, N_2. There is one extra pair of σ bonding electrons and two extra pairs of p bonding electrons. N_2 has a triple bond.*

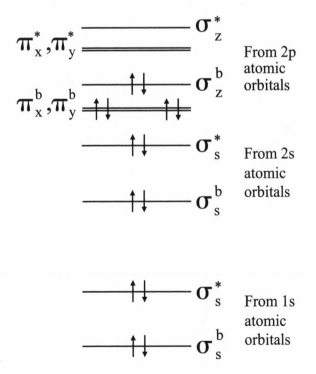

electrons each carbon needs to obtain the Ne configuration. Carbon needs to make four bonds to obtain the Ne configuration by forming molecules such as CH_4. It can't make four bonds by forming the C_2 molecule, and C_2 doesn't exist.

F_2 has a single bond, O_2 has a double bond, and N_2 has a triple bond. Table 13.1 illustrates how the bond order strongly affects the properties of the bond. As the bond order increases, the bond length gets shorter and the bond energy increases. The bond energy is the energy that needs to be put into the molecule to break the bond. Breaking the bond means separating the atoms to a distance so far apart that they no longer feel each other. As will be discussed in the next chapter, carbon can make single, double, and triple bonds to another carbon atom while at the same time forming bonds to other atoms, such has hydrogen. Before we can discuss molecules that are larger than diatomics, we need to go beyond homonuclear diatomic molecules and examine heteronuclear diatomics to see how molecular orbitals are formed from nonidentical atoms.

HETERONUCLEAR DIATOMICS

In homonuclear diatomics, the MOs are formed from atomic orbitals with identical energies. In a heteronuclear diatomic, for example hydrogen fluoride (HF), the two atoms are different. Because the atoms are different, the atomic orbitals' energies of one atom will

TABLE 13.1. *The Effect of Bond Order on Bond Properties.*

	F_2	O_2	N_2
bond order	single (1)	double (2)	triple (3)
bond length (Å)	1.42	1.21	1.10
bond energy (10^{-19} J)	2.6	8.3	15.6

not match those of the other. For HF, the hydrogen atom has a single electron in the 1s orbital. F has nine electrons in the 1s, 2s, and 2p orbitals. Both F_2 and H_2 have single bonds. Looking at Figure 13.6, the single bond in F_2 is a σ bond due to the bonding MO, σ_z^b . This bonding MO was formed from two $2p_z$ atomic orbitals, one on each F atom. H_2 has a single σ bond due to the bonding MO formed from two 1s orbitals (see Figure 12.7). To make HF, the question is, which orbital on F will combine with the 1s orbital on H to give MOs involved in bonding? Quantum theory calculations show that states (atomic orbitals) that are close in energy can combine to produce electron sharing MOs. Atomic orbitals with significantly different energies form MOs that are essentially equal to the individual atomic orbitals, and they do not contribute to bonding.

The hydrogen 1s orbital has an energy of -2.2×10^{-18} J. (Recall that the negative sign means that the electron is bound.) The energy of the fluorine 1s orbital (measured in the F_2 molecule) is -1.1×10^{-16} J. So the 1s orbital of F is about 50 times lower in energy than the 1s orbital of H. This is a tremendous difference in energy, and the H 1s will not form a MO with the F 1s. In contrast, an F 2p orbital has an energy of -2.8×10^{-18} J. This energy is about 25% below the H 1s energy, which is close in energy. So an F 2p and the H 1s orbital energies are similar enough that they will form true MOs.

There are three F 2p orbitals, $2p_z$, $2p_x$, and $2p_y$. To decide which of these will interact with the H 1s, we need to define how we bring the atoms together. Let's say we bring the H atom and the F atom toward each other along the z axis as shown at the top of Figure 13.10. The two circles reflect the correct relative sizes of the H and F atoms. Fluorine's $2p_y$ orbital has its lobes perpendicular to the z axis, as shown in the middle portion of the figure. (The orbitals are not drawn to scale.) When the $2p_y$ overlaps the hydrogen 1s orbital, the positive $2p_y$ lobe will constructively interfere with the 1s orbital, but the negative lobe will destructively interfere. The net result is

FIGURE **13.10.** *Top: H and F atoms brought together along the z axis. Circles show the relative size of the atoms. Middle: overlap of the H 1s and F 2p_y orbitals. There is equal constructive (+) and destructive (−) interference in the overlap region. No MO formation. Bottom: overlap of the H 1s and the F 2p_z orbitals. There is constructive interference in the overlap region.*

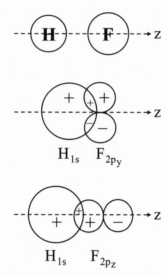

that there is no net constructive or destructive interference. The same is true for the $2p_x$ orbital. The $2p_y$ and $2p_x$ orbitals will not form bonding or antibonding MOs in the HF molecule. The lower portion of the figure shows the $2p_z$ orbital's positive lobe overlapping with the 1s orbital, which is also taken to be positive. This overlap will result in constructive interference of the probability amplitude waves, and can produce a bonding MO. Since there is electron density along the line connecting the nuclei, the bond will be a σ bond. If the F negative $2p_z$ lobe overlaps with the H positive 1s lobe, there will be destructive interference that gives rise to an antibonding MO.

As discussed above, the fluorine 1s orbital energy is so much lower than the hydrogen 1s, that the fluorine 1s electrons do not participate in bonding. The outermost electrons in an atom, that is, the last shell that is filled, are the ones that contribute to bonding. These are the valence electrons. For elements in the second row of the Periodic Table, like fluorine, the 2s and 2p are the valence electrons. In making molecular orbital energy level diagrams, usually only the orbitals associated with the valence electrons are shown because these are the orbitals that can be involved in bonding. Figure 13.11 shows the MO energy level diagram for HF with the F 1s orbital and electrons left out. The energy level spacings are not to scale. As discussed in connection with Figure 13.10, the H 1s atomic orbital will combine with the F $2p_z$ atomic orbital to form bonding (σ^b) and antibonding (σ^*) MOs. This is indicated in the diagram by the dashed lines. This diagram is similar to the energy level diagram in Figure 13.5 except that now the atomic orbitals that form the MOs do not have identical energies.

Fluorine has nine electrons. Two are in the 1s orbital, which leaves seven. Hydrogen has one electron. So there are a total of eight valence electrons to place in the MO energy levels. The first two go into the level labeled 2s. The fluorine 2s orbital is much lower in energy than the hydrogen 1s, and these electrons do not participate in bonding. Therefore, the 2s molecular orbital is essentially the same as the fluorine 2s atomic orbital. The two electrons in the 2s orbital are a lone pair. The next two electrons go into the σ^b bonding MO. The final four electrons go into the $2p_x$ and $2p_y$ orbitals. Again, these are basically atomic orbitals of fluorine. They do not play a role in the bonding. These four electrons comprise two more lone pairs. While the lone pairs do not play a role in bonding, in polyatomic molecules they influence the shapes of molecules, which will be discussed in Chapter 14. The net result is that there are two electrons in the bonding MO and none in an antibonding MO. Therefore, HF has a single bond. Hydrogen and

FIGURE **13.11.** *Molecular orbital energy level diagram for HF. The atomic orbitals of the valence electrons are shown at the right and left. The F $2p_z$ atomic orbital combines with the H 1s atomic orbital to give a bonding (σ^b) and antibonding (σ^*) MO. σ^b is filled with one H electron and one F electron. σ^* is unfilled. The net is one bond. The other F electrons do not participate in the bonding. They are lone pairs of electrons.*

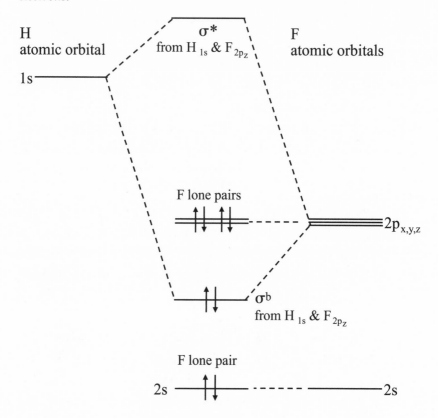

fluorine share a pair of electrons in the bonding MO. For H, the sharing provides the additional electron necessary to achieve the helium rare gas configuration. For F, the sharing provides the extra electron needed to obtain the neon rare gas configuration.

VISUAL MODELS OF MOLECULES

HF like F_2, O_2, and N_2 are diatomics and therefore linear molecules. In the next chapter, we will talk about molecules with more complex shapes. The structure of molecules can be shown in a number of ways. HF can be written as H − F to indicate there is a single bond. In more complicated molecules, this type of representation can show which atoms are bonded to each other and the bond order. However, it cannot display the three-dimensional geometry or give a feel for what the molecule actually looks like. Now to say a molecule looks like something is fundamentally incorrect. HF has two nuclei surrounded by the probability amplitude waves that are the electrons. Nonetheless, there are representations that are useful in discussing the nature of molecules. Figure 13.12 displays two such representations of HF. The top portion is a ball-and-stick molecular model. It shows the connection between the atoms and their relative size. H is light in color and F is dark in color. The bond between

FIGURE 13.12. *Representations of the HF molecule. H: light; F: dark. Top: a ball-and-stick version that shows how the atoms are bonded and the relative sizes of the atoms. Bottom: a space-filling version that is more realistic.*

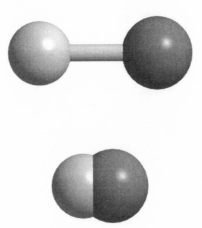

the atoms is much too long. The bottom portion of the figure shows a space-filling model. Most of the electron density is inside of the overlapping spheres. The sizes and the internuclear separations are correct. The colors and the sharp line between the atoms are to aid the eye. There is no actual separation of the electrons between the atoms.

The material presented in this chapter and in the next chapter is necessary to understand bonding in polyatomic molecules. In the next chapter, we need to extend the ideas presented here to molecules with more than two atoms. Polyatomic molecules have shapes, and to understand the shapes, we will introduce new ideas, hybrid atomic orbitals. The material developed in Chapters 13 and 14 will be used in subsequent chapters to examine a wide variety of problems such as what are trans fats and why are they different from other fats.

14

Bigger Molecules: The Shapes of Polyatomic Molecules

THE WORLD AROUND US is composed of polyatomic molecules. Polyatomic molecules are molecules with more than two atoms. These range in size from triatomic molecules, such as carbon dioxide (CO_2), which is a major greenhouse gas, to molecules with thousands of atoms, such as large proteins that are responsible for most biological functions. As discussed in Chapter 13, diatomic molecules can only have one shape, linear. Larger molecules, however, can have very complicated shapes and structures. For example, saturated fats, unsaturated fats, or polyunsaturated fats differ by their shapes and structures, which are determined by the nature of chemical bonds. A given large molecule can have more than one shape. Trans fats, which are currently being at least partially removed from food (see Chapter 16), only differ in their shape from molecules made up of the same sequence of atoms that are not trans fats. A central question in molecular matter is, how do molecules achieve their shapes and how are different shapes possible for molecules made of the same atoms connected together in the same way?

Before proceeding, it is worth stating that the covalent bond, which is responsible for holding atoms together to form molecules, is an intrinsically quantum mechanical phenomenon. It was not possible to explain the nature of chemical bonds or the structure of molecules before the advent of quantum theory. Linus Pauling (1901–1994) won the Nobel Prize in Chemistry in 1954 "for his research into the nature of the chemical bond and its application to the elucidation of the structure of complex substances." It is note-worthy that Linus Pauling is one of the few people to win two Nobel Prizes. In 1962 he won the Nobel Peace Prize.

MOLECULAR SHAPES— TETRAHEDRAL METHANE

To examine the shapes of molecules and how chemical bonding can give rise to different shapes, we need to introduce some new ideas about atomic orbitals. We will use methane as an example of a rela-tively simple polyatomic molecule to bring out the important issues.

Methane (natural gas) is CH_4. Now, look again at carbon's posi-tion in the Periodic Table (Figure 11.4). Note that carbon needed to make four electron pair sharing covalent bonds to achieve the neon closed shell configuration. In methane, carbon makes four bonds with four hydrogen atoms. The connectivity of the atoms can be shown in a simple diagram, which is the left portion of Figure 14.1. Each line represents an electron pair covalent bond. However, this diagram does not tell much about methane's shape. Methane is not flat. The right side of the figure shows a ball-and-stick model of the three-dimensional shape of methane. (See the end of Chapter 13 for a discussion of ball-and-stick and space-filling models.) Methane is a perfect tetrahedron. Imagine having a model like the right side of Figure 14.1 and gluing a triangular piece of paper with sides of equal length that just covered three of the hydrogen atoms. You can glue on four such triangles, three sides and the bottom. These four trian-gles form a perfect triangular pyramid with the hydrogens at the

FIGURE 14.1. *Left: diagram showing the bond connectivity of methane, but not its three-dimensional shape. Right: a three-dimensional ball-and-stick model of methane that shows the tetrahedral shape of the molecule.*

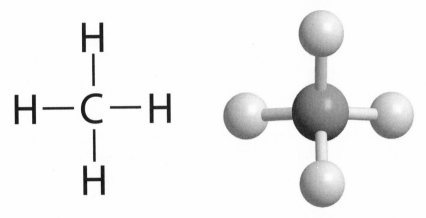

four apices and the carbon in the middle. The angle formed by the line from a hydrogen to the carbon and then a line from the carbon to another hydrogen is exactly 109.5°. This is true of all four angles. They are identical. The angle for a perfect tetrahedral molecule is 109.5°.

Minimizing Repulsion Between Bonds Determines the Shape

Why does methane have a tetrahedral shape? In Chapter 13, we saw that bonding molecular orbitals concentrate electron density between the atomic nuclei. The concentration of electron density between the nuclei is shown in Figures 13.2 and 13.3 for σ and π bonds. Chapter 13 discussed diatomic molecules in which only two atoms are bonded. We did not have to worry about how multiple atoms with some set of bonds connecting them would be arranged. While quantum theory can calculate the details of the shapes of molecules, the basic reason why a molecule will have a particular shape, such as tetrahedral, is quite simple. In polyatomic molecules, sharing electrons between two atomic nuclei to form a bond concentrates electron density between the nuclei, just as it does with diatomic molecules. However, in polyatomics, we have a number of bonds, each of which has a high density of negatively charged electrons. The negatively charged regions, the bonds, repel each other. Basically, the bonds want to stay as far away from each other as possible. These regions of high negatively charged electron density want to overlap as little as possible. Forming bonds lowers the energy of the system relative to the separated atoms. If the energy is not lowered, bonds will not be formed. But to make the energy as low as possible, the system of atoms assumes a configuration that minimizes the electron repulsions by keeping the bonds away from each other.

In methane, the tetrahedral shape minimizes the repulsions between the bonds. Look at the left-hand drawing in Figure 14.1. The four hydrogens are in a plane. If I keep them in the plane, they are as well separated as possible. The angle formed by two adjacent bonds is 90°. With the atoms in the plane, if I make one of the angles larger than 90° to further separate two of the hydrogen-carbon bonds, these bonds get closer to the other two bonds. So if the four hydrogen atoms are kept in the same plane as the carbon atom, the best that can be done is to have 90° bond angles.

However, there is no reason that all of the atoms have to lie in a plane. In the structure on the left side of Figure 14.1, the angle between the top and bottom H atoms is 180°, and these two C-H bonds are far apart; the same is true for the left and right H atoms. Imagine pulling the top and bottom H atoms above the plane of the page, while keeping the bond lengths the same and pushing the right and left H atoms below the plane of the page. The top and bottom C-H bonds are getting closer together and the angle is reduced to less than 180°, but they are getting further away from the left and right C-H bonds. Bringing the top and bottom C-H bonds above the plane and pushing the left and right C-H bonds below the plane reduces the overall bond-bond interaction. The interaction between the top and bottom bonds is increased, but they were very far apart to begin with. This is also true of the left and right bonds. However, if the top-bottom and left-right angles are reduced too much, the repulsion will go up again. There is a best angle, and that is 109.5°, the tetrahedral angle. This is the angle that keeps the electrons in the four C-H bonds as far apart as possible.

Lone Pairs Also Matter

In Chapter 11, we found that to obtain a closed shell configuration, C needs to form four bonds, N needs to form three bonds, and O needs to form two bonds. If the bonds are to hydrogens, then we have methane, ammonia, and water, CH_4, NH_3, and H_2O. In discussing HF at the end of Chapter 13, we noted that some of the F electrons were not involved in bonding at all. The electrons were paired in what are essentially atomic orbitals called lone pairs. Lone pairs are pairs of nonbonding electrons that give rise to high electron density in the region of space they occupy. Electrons in bonds do not want to be near electrons in lone pairs. So, although they are not bonds, lone pairs also influence the shapes of molecules. In a bond, the electrons are shared and more or less concentrated between two atoms. Lone pairs do not have a second atom to hang on to. As a result, the lone pair electron distribution is bunched closer to the atom they belong to, and it is somewhat "fatter" than a bond pair distribution.

Figure 14.2 shows models of methane, ammonia, and water. Ammonia has one lone pair, and water has two lone pairs. If you include the lone pairs, all three molecules are basically tetrahedral in shape. However, ammonia and water are not perfect tetrahedrons. The lone pair in ammonia is more spread out than the bond pairs. To minimize the overall electron-electron repulsion to produce the lowest energy, the bonds move away from the lone pair, thereby bringing the bonds closer together. The HNH angle in am-

Figure 14.2. *Methane (left), ammonia (center), and water (right). Lone pair electrons repel the bond pair electrons, pushing the bonds closer together, which reduces the angle between bonds to H atoms from the central atom.*

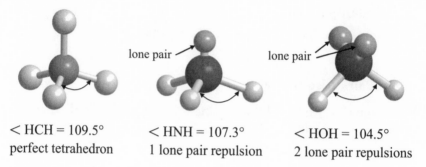

< HCH = 109.5°   < HNH = 107.3°   < HOH = 104.5°
perfect tetrahedron   1 lone pair repulsion   2 lone pair repulsions

monia is 107.3°, slightly less than the perfect tetrahedral angle. Water has two lone pairs, causing the angle between the hydrogen-oxygen bonds to be reduced further to 104.5°.

Trigonal-Shaped Molecules

If a central atom is bound to only three other atoms, it will have a trigonal shape with the four atoms lying in a plane. Figure 14.3 shows two trigonal molecules, BH_3 and H_2CO (formaldehyde). BH_3 exists, but it is very reactive because it is two electrons short of the neon closed shell configuration. In BH_3, each H has a single bond to the B. The HBH angle is exactly 120°. The hydrogens form a perfect equilateral triangle. This is the shape that keeps the bonds as far apart as possible, which lowers the energy by reducing the repulsive interactions between the bonding electrons in each bond.

In Chapter 13, the MO diagram for O_2 (Figure 13.8) showed that the oxygen molecule has a double bond. In formaldehyde (the smelly liquid in the jars containing dead things in biology class), the O is double bonded to the C. The double bond is shown in the

FIGURE **14.3.** *Left: BH_3. The atoms lie in a plane. The HB bonds are single, and the hydrogens form a perfect equilateral triangle. Each HBH bond angle is 120. Right: H_2CO (formaldehyde). The atoms lie in a plane. The CO bond is a double bond. The angles are unequal.*

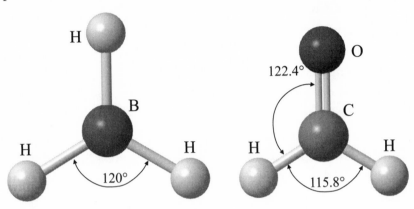

ball-and-stick figure model as two cylinders joining the atoms rather than one. The double bond gives the O the neon closed shell configuration as in the O_2 molecule. The C needs to share two additional electrons to obtain the neon closed shell configuration, which it does by single bonding to two H atoms. We will discuss double bonds in detail to see how they can be formed from atomic orbitals. Here, we only need to recognize that a double bond concentrates two pairs of electrons between the C and the O. Because of the extra electron density, a double bond is fatter than a single bond. The fatter $C=O$ double bond pushes the $C-H$ single bonds away from it and toward each other. The angles are shown in Figure 14.3. The result is that formaldehyde is still a planar trigonal molecule, but it is not a perfect equilateral triangle.

PROMOTING ELECTRONS

Returning to methane, the question is how does methane make four tetrahedrally configured bonds? In Chapter 11, we discussed the electronic configuration of the atoms (see Figure 11.1). Carbon has six electrons, two in the 1s, two in the 2s, and two in 2p orbitals. The valence electrons, the electrons used in bonding, are the 2s and 2p electrons. The top portion of Figure 14.4 shows the atomic orbital energy levels with the four valence electrons filled in. The 1s electrons are not shown. As discussed in Chapter 11 and earlier in this chapter, carbon will form four bonds. In methane, it forms four electron pair sharing bonds to four hydrogen atoms. Each H atom contributes one electron. So the carbon must have four unpaired electrons to form bonds. Each carbon unpaired electron can join with one electron from an H atom to yield an electron pair bond. To have four unpaired electrons, carbon "promotes" a 2s electron to a 2p orbital, as shown in the bottom part of Figure 14.4. For an isolated carbon atom, the configuration shown in the bottom of the figure would not occur unless a lot of energy was pumped into the

FIGURE 14.4. *Top: Atomic carbon valence orbitals with the four valence electrons. Bottom: When bonding, a carbon atom "promotes" a 2s to a 2p electron to give four unpaired electrons used to form four bonds to other atoms.*

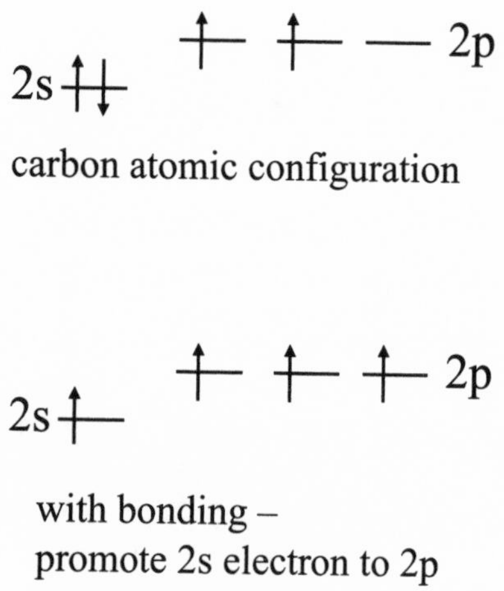

$$2s \uparrow\downarrow \qquad \uparrow \quad \uparrow \quad \underline{} \quad 2p$$

carbon atomic configuration

$$2s \uparrow \qquad \uparrow \quad \uparrow \quad \uparrow \quad 2p$$

with bonding –
promote 2s electron to 2p

atom. For a carbon atom, putting a 2s electron into a 2p orbital is a high energy configuration. However, when atoms form molecules, the electrons and nuclei of the different atoms affect each other. Imagine bringing four H atoms toward a C atom. Now, the system wants to assume the lowest energy configuration for all five atoms. Forming the four bonds lowers the energy more than putting a 2s electron into a 2p orbital raises the energy.

HYBRID ATOMIC ORBITALS— LINEAR MOLECULES

So we see how carbon can make four bonds to yield methane, but why is the shape tetrahedral? The three 2p orbitals are p_x, p_y, and p_z. These three orbitals are perpendicular to each other, with a 90° angle between any pair. If three H atoms bonded to the three 2p orbitals, the bond angles would be 90°. Furthermore, the 2s orbital is spherical. The fourth H atom's 1s orbital would have to combine with the carbon 2s orbital. Without something else happening, it is pretty clear that using the carbon 2s and the three 2p orbitals is not going to give methane four identical C-H bonds in a tetrahedral configuration. In addition, how does carbon make the trigonal molecule formaldehyde, or the linear molecule carbon dioxide ($O = C = O$). In each case, tetrahedral, trigonal, or linear, carbon bonding involves the same 2s and 2p orbitals.

Formaldehyde and carbon dioxide involve double bonds, which we will get to shortly. To bring out the important features of the nature of atomic orbitals that can give linear, trigonal, and tetrahedral shapes, we will examine bonding in BeH_2, BH_3, and CH_4, that is, beryllium dehydrate, boron trihydride (borane), and methane. Be and B in BeH_2 and BH_3 do not have closed shell neon rare gas configurations. Therefore, they are very reactive. These molecules can be made, but they will react with virtually anything they come in contact with to make new molecules in which the Be and B have closed shell configurations. Here we will use them as convenient examples.

Be is two electrons past the He closed shell configuration. As an atom, these two electrons would be paired in the 2s orbital. These are Be's valence electrons. In BeH_2, each H has one electron in a 1s orbital. For Be to make two electron pair bonds, one to each H atom, it will promote one of the 2s electrons to a 2p orbital, which we will take to be the $2p_z$, as shown at the top of Figure 14.5. The second line of the figure shows schematics of the 2s and the $2p_z$ orbitals separated. These actually have the same center, the nucleus of the Be. These orbitals are electron probability amplitude waves. Waves can be added and subtracted to give new waves. We start with two atomic orbitals, the 2s and the $2p_z$, and by addition and subtraction, we get two new atomic orbitals, called hybrid orbitals. When we add the waves together, we get regions of constructive interference and regions of destructive interference because the lobes of the probability amplitude waves have signs. The third line in Figure 14.5 shows the sum of the 2s and $2p_z$ orbitals. This is called an sp hybrid, which we have labeled sp_{z+} because it is an sp hybrid made from an s orbital and the $2p_z$ orbital, and its big positive lobe points long the $+z$ direction. The second to the bottom line of Figure 14.5 shows the 2s minus the $2p_z$. One way to think of this is just that we have flipped the $2p_z$ orbital shown in the top line of the figure around so the positive lobe points to the left rather than to the right, and then we add. We have labeled this hybrid orbital sp_{z-} because its big positive lobe points in the $-z$ direction. This orbital has the same shape as the sp_{z+} orbital but it points in the opposite direction. We start with two atomic orbitals, 2s and $2p_z$, and we end with two hybrid atomic orbitals with their positive lobes pointing along the z axis in opposite directions. The bottom line of Figure 14.5 shows the schematic of both hybrid atomic orbitals as they would occur in the Be atom. There is one Be nucleus with the two hybrid orbitals pointing along the $+z$ and $-z$ directions.

FIGURE 14.5. *Top: Be valence electrons with one promoted to the $2p_z$ orbital. Next: The 2s and $2p_z$ orbitals of Be shown separated. Next: The sum of 2s and $2p_z$ to form the hybrid atomic orbital sp_{z+}. Next: $2s - 2p_z$ to form the hybrid orbital sp_{z-}. Bottom: The two hybrid orbitals of Be point in opposite directions along z.*

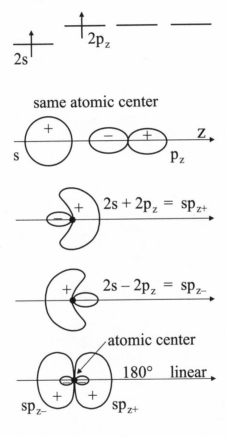

To make BeH_2, Be will use the two hybrid atomic orbitals to form electron pair bonding molecular orbitals with two hydrogens. The bonding is shown schematically in Figure 14.6. The top of the figure shows two hydrogen atoms, H_a and H_b, approaching a Be. Hydrogen atom electrons are in 1s orbitals, $1s_a$ and $1s_b$. The beryllium has its two hybrid atomic orbitals, sp_{z-} and sp_{z+}, pointing at the hydrogen 1s orbitals. The middle portion of the figure shows a

FIGURE 14.6. *Top: Two H atoms approach a Be. Middle: The H 1s orbitals form electron pair bonds with the two Be sp hybrid orbitals to produce the linear BeH_2 molecular shown as a ball and stick model at the bottom.*

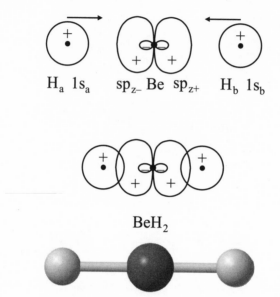

schematic illustration of the overlapping atomic orbitals. The hydrogen $1s_a$ orbital on the left side of the figure will form a bonding MO with the Be sp_{z-} hybrid atomic orbital. This bonding MO will contain two electrons, one from the hydrogen and one of the two Be valence electrons. The hydrogen $1s_b$ orbital on the right side of the figure will form a bonding MO with the sp_{z+} hybrid orbital. The electron from H_b and the other Be valence electron will form another covalent bond. These are σ bonds since there is electron density along the line connecting the nuclei. The result is the linear BeH_2 molecule shown at the bottom of the figure.

In total the BeH_2 molecule has six electrons counting the two pairs of bonding electrons and the two electrons in the Be 1s orbital.

It is important to note that all of these electrons are probability amplitude waves that span the entire molecule. When we say that the $1s_a$ electron makes an electron pair bond with the sp_{z-} electron, this is bookkeeping. All of the electrons are free to roam over the entire molecule. However, the probability distributions for the electrons are such that at any given instant there is electron density associated with the Be and one of the H atoms that corresponds to one bond and electron density associated with the Be and the other H atom that corresponds to the other bond.

HYBRID ATOMIC ORBITALS— TRIGONAL MOLECULES

As discussed in connection with Figure 14.3, BH_3 is a trigonal molecule with 120° between bonds. Atomic boron has three valence electrons, two in the 2s and one in a 2p orbital. To form three electron pair sharing bonds to three hydrogen atoms, the boron atom needs three unpaired electrons. As shown at the top of Figure 14.7, B will promote one electron from a 2s orbital to a 2p orbital to have the three unpaired electrons. If the molecule lies in the xy plane, then the 2p orbitals involved in bonding will be the $2p_x$ and $2p_y$. To form the equilateral trigonal BH_3 molecule, the three boron atomic orbitals will hybridize to yield the three hybrid atomic orbitals, sp_a^2, sp_b^2, and sp_c^2. The sp^2 notation means that the hybrid orbital is composed of an s orbital and two different p orbitals. We start with three different orbitals, 2s, $2p_x$, and $2p_y$. Orbitals are never gained or lost, so we end up with three different hybrid orbitals. These are shown in the middle portion of Figure 14.7. Each orbital has a positive and a negative lobe. The adjacent lobes have an angle of 120° between them. Each of these contains one of the three unpaired B valence electrons.

The bottom portion of Figure 14.7 shows the bonding of B with three H atoms. Each H atom has a single 1s electron. An H 1s

FIGURE 14.7. *Top: B valence electrons with one promoted to a 2p orbital. Middle: the 2s, 2p$_x$, and 2p$_y$ orbitals of B combine in three combinations to form three hybrid atomic orbitals, sp$_a^2$, sp$_b^2$ and sp$_c^2$. The angle between the lobes is 120. Bottom: The three B hybrid orbitals forming bonds with the three H atom 1s orbitals.*

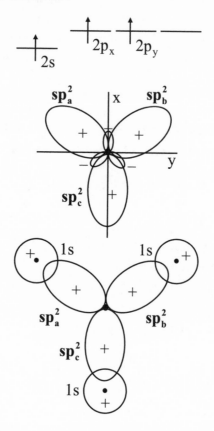

orbital combines with a B sp^2 orbital to form a bonding molecular orbital. There are two electrons in this bonding MO, one from the H and one from the B. The result is an electron pair bond. Each of the bonds between B and H is a σ bond since there is electron density along the line connecting the nuclei. A model of BH$_3$ is shown in Figure 14.3.

HYBRID ATOMIC ORBITALS— TETRAHEDRAL MOLECULES

In methane, carbon makes four bonds to four hydrogen atoms. As discussed above and shown in Figures 14.1 and 14.2, methane is tetrahedral. As illustrated in Figure 14.4, to make four electron pair sharing covalent bonds, carbon promotes one of its 2s electrons to a 2p orbital. It then has four unpaired electrons in the 2s, $2p_x$, $2p_y$, and $2p_z$ orbitals. As discussed in some detail following Figure 14.4, these four carbon atomic orbitals will not yield four identical bonds to four hydrogens and produce a tetrahedral molecule. Therefore, the 2s, $2p_x$, $2p_y$, and $2p_z$ orbitals combine in four different combinations to form four atomic hybrid orbitals, sp_a^3, sp_b^3, sp_c^3, and sp_d^3. The sp^3 designation means that each of the four hybrid atomic orbitals is a combination of an s orbital and the three different p orbitals. Figure 14.8 shows the four sp^3 orbitals overlapping with the four hydrogen 1s orbitals. Only the positive lobes of the sp^3 orbitals are shown. Each of them has a small negative lobe pointing in the opposite direction from the positive lobe in a manner analogous to that shown for the sp^2 orbitals in the middle portion of Figure 14.7. The orbitals represented by dashed curves are in the plane of the page. The orbital shown as a solid curve is coming out of the plane of the page at an angle, and the orbital shown as the dot-dash curve is at an angle going into the plane of the page. The angle between any pair of the sp^3 lobes is 109.5°, giving the perfect tetrahedral shape discussed in connection with Figures 14.1 and 14.2.

Figure 14.2 shows methane, ammonia, and water. We said that each of these is tetrahedral when the lone pairs are included, but ammonia and water are slightly distorted tetrahedrons. Like methane, ammonia and water also use sp^3 hybridization to form bonds. The nitrogen in ammonia, NH_3, has five valence electrons. Two of them form a lone pair. The lone pair does not participate in bonding. The nitrogen uses three of its four sp^3 hybrid orbitals to bond

FIGURE **14.8.** *Four carbon sp³ hybrid atomic orbitals and four hydro-gen 1s orbitals for carbon bonding to four hydrogens in methane. Dashed orbitals in the plane of the page. Solid orbitals out of the plane. Dot-dash orbitals into the plane. Only the positive lobes of the sp³ hybrids are shown. The four sp³ hybrids form a perfect tetrahe-dron.*

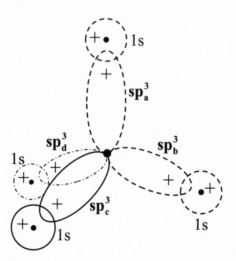

to the three H atoms. The fourth sp³ orbital contains the lone pair. As discussed, the H-N-H bond angle is a little less than the perfect tetrahedral angle, 109.5°, because the spatial distribution of the lone pair electrons is somewhat fatter than the N-H bond pair electrons, and the fatter lone pair pushes the N-H bonds slightly toward each other. The oxygen in water, H_2O, has six valence electrons. Four of them form two lone pairs. These two lone pairs of electrons do not participate in bonding. Oxygen uses two of its sp³ hybrid orbitals to form bonds to the two hydrogens. The other two sp³ orbitals are each occupied by a lone pair. These lone pairs cause the H-O-H angle to be less than the perfect tetrahedral angle of 109.5° (see Figure 14.2).

HYDROCARBONS WITH SINGLE BOND

A hydrocarbon is a molecule that is solely made up of carbon and hydrogen atoms. First we will discuss more complicated hydrocarbons than methane, but initially only molecules with single bonds. The next simplest hydrocarbon after methane is ethane. Ethane has two carbons and six hydrogens with the chemical formula, C_2H_6. Figure 14.9 displays the ethane structure in three ways. The top shows only the bonds between the atoms. Each carbon has a single bond to three hydrogens and a single bond to the other carbon. The middle portion of the figure shows the hybrid atomic orbitals used in bonding. To make the four electron pair sharing bonds, each carbon uses four sp^3 hybrid orbitals, just like in methane. Three of the orbitals on each carbon are used to make bonds to three hydrogens. A sp^3 orbital combines with a hydrogen 1s orbital to form a bonding MO to yield a σ bond. The fourth sp^3 orbital belonging to one carbon forms an MO with the sp^3 orbital on the other carbon to produce a carbon-carbon σ bonding MO.

The bottom portion of Figure 14.9 is a standard method of schematically displaying the spatial arrangement of atoms in a molecule. The atoms that are bonded and lie in the plane of the page are connected with lines. Bonded atoms that stick out of the plane of the page are connected by narrow filled triangles with the sharp end of the triangle pointing at the atom in the plane and the base of the triangle next to the atom out of the plane. Bonded atoms that stick into the plane of the page are connected by open triangles with the base of the triangle next to the atom in the plane and the point of the triangle next to the atom below the plane. As shown in the bottom portion of the figure, each carbon is at the center of a tetrahedron. Any pair of bonds around a carbon has the tetrahedral angle of 109.5° between them. The C-H bond length is 1.07 Å (1.07 $\times$ 10^{-10} m), and the C-C bond length is 1.54 Å. Carbon atoms are

FIGURE **14.9.** *Three diagrams of ethane,* C_2H_6. *Top: The bonds between atoms. Middle: Each carbon has four sp^3 hybrid atomic orbitals, three bond to hydrogens and the fourth to the other carbon. Solid lines: in the plane of the page; dashed lines: out of the plane; dotted lines: into the plane. Bottom: Method for showing spatial arrangement. Lines: in the plane of the page; filled triangles stick out of the page; open triangles stick into the page. Each carbon's bonds are tetrahedral, with bond angles 109.5°. The C-C bond is longer than the C-H bonds.*

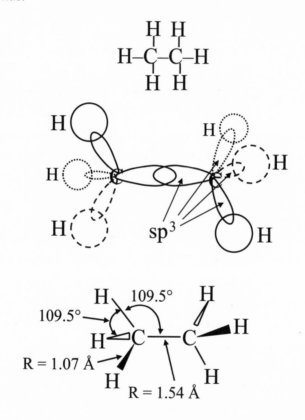

bigger than hydrogen atoms, so the separation of the atomic centers is greater for the two carbon atoms than for a C and an H.

Figure 14.10 shows a ball-and-stick model (top) and a space filling model (bottom) of ethane. In Figures 14.9 and 14.10, we have

FIGURE 14.10. *Ethane ball-and-stick model (top), space-filling model (bottom). The atoms are the same size in the two models.*

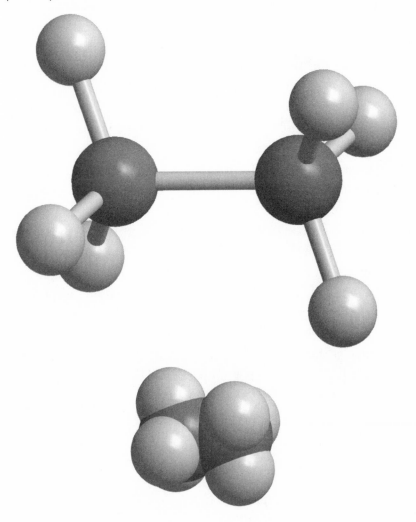

five different representations of the ethane molecule. Only the spacing-filling molecule gives a real feel for the spatial nature of the
molecule. The other four representations exaggerate the separations
between the atoms for clarity. The atoms in the ball-and-stick and
spacing-filling models in Figure 14.10 are the same sizes. In the
ball-and-stick model, the bonds are shown as cylinders, and the
atoms are separated by the bonds. It is important to recognize that
the bonding comes from the formation of molecular orbitals. The
electrons are shared by the atoms, which are not separated as in the
ball-and-stick model or the other representations. The surface in the
space-filling model contains a large fraction of the electron probability distribution. In the space-filling model, the atoms are shaded
differently so that they can be distinguished.

We need to discuss one more relatively simple molecule, propane, before the molecules get large enough that some general features can be brought out. Propane has three carbon atoms and eight
hydrogens. Its chemical formula is C_3H_8. The formula doesn't tell
how the atoms are connected. It can also be written as H_3C-H_2C-
CH_3. In this notation, it is understood that the hydrogens are
bonded to a carbon. The carbons are bonded to each other with
single bonds. The end carbons are bonded to three hydrogens and
another carbon. The center carbon is bonded to two hydrogens and
two carbons. Figure 14.11 shows two representations of propane.
The top diagram shows the bonding and angles between the bonds.
Each carbon uses four sp^3 hybrid orbitals to make four σ bonds.
The carbons are tetrahedral, with the C-C-C angle 109.5° and the
HCH angles 109.5°. The bottom portion of the figure shows a ball-
and-stick model of propane.

LARGE HYDROCARBONS HAVE
MULTIPLE STRUCTURES

Methane, ethane, and propane have only one way that the atoms
can be bonded together and only on spatial conformation. Butane

FIGURE 14.11. *A diagram and a ball-and-stick model of propane, C_3H_8.*
The carbon centers are tetrahedral.

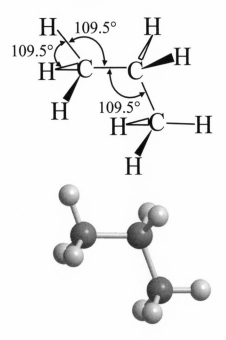

and all larger hydrocarbons have multiple structural configurations
(the way the atoms are bonded together) and more than one spatial
configuration for a particular structural configuration. Butane has
four carbons. Its chemical formula is C_4H_{10}. There are two distinct
structural forms of butane. These are called structural isomers. Figure 14.12 displays the two structural isomers of butane. Both molecules have the same number of carbons and hydrogens, but they
have very different shapes. Butane can be n-butane, which is normal butane. If we take propane and add one more carbon on the
end, we get n-butane. n-butane is referred to as a linear chain because a carbon is at most bonded to two other carbons, one on
either side. As can be seen in the ball-and-stick model, the molecule
is not actually linear because each carbon has a tetrahedral arrangement of bonds formed using four sp³ hybrid orbitals.

FIGURE 14.12. *Two structural isomers of butane, C_4H_{10}. In the diagram at the top, CH_3 represents a carbon bonded to three hydrogens. N-butane is a linear chain in the sense that each carbon is bonded to at most two other carbons. Isobutane is branched. The central carbon is bonded to three other carbons.*

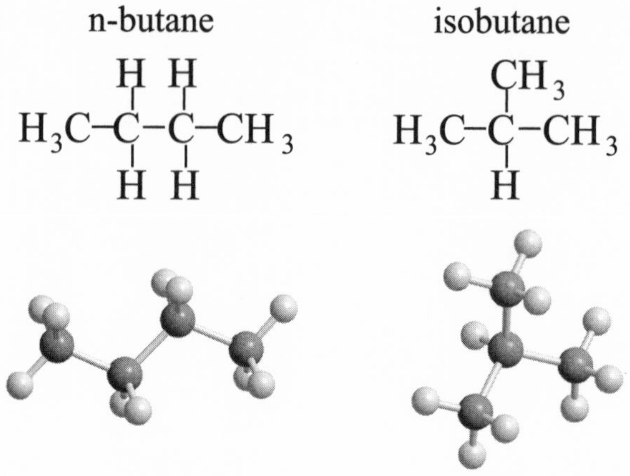

As shown in Figure 14.12, butane can have another isomer, called isobutane. Isobutane has a central carbon connected to three other carbons and one hydrogen, with each of the other three carbons only connected to the central carbon and to three hydrogens. All four carbons use sp³ hybrid atomic orbitals for bonding and are tetrahedral. Isobutane is referred to as branched. The fact that butane can have the same number of carbons and hydrogens but two different structures is of great importance. Molecules with more carbon atoms can have many more than two structures.

In addition to the two structural isomers, n-butane has two conformers. Conformers are different shapes, conformations, for the same set of atoms that are connected in the same way. They differ because it is possible for rotation to occur about a C-C single bond. Figure 14.13 shows n-butane in two conformations called trans and gauche. Both conformers shown in the figure are n-butane because

FIGURE 14.13. *Two conformers of n-butane. The gauche form is obtained from the trans form by a 120° rotation about the center C-C bond.*

n-butane

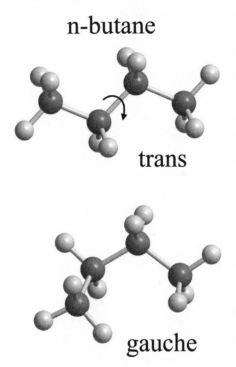

trans

gauche

the carbon atoms are connected in the same way. If you take the top conformer and rotate the central carbon-carbon bond, as shown by the arrow, 120°, you get the gauche form. The trans conformer has all of the carbons in a plane. The gauche form has three of the carbons in a plane and the fourth carbon is sticking out of the plane of the page. There is actually another gauche form that comes from rotating the trans form around the central C-C bond 120° in the opposite direction from that shown by the arrow. In that case the same three carbons are still in the plane, but the fourth carbon is sticking into the page instead of out of the page. These two gauche forms in some sense have the same shape, but they are not the same. They are like a left hand and a right hand glove. Like the gloves, these two gauche forms cannot be superimposed on each other. They are the mirror images of each other. A carbon center that can have a right hand and left hand form, depending on a rotation, is said to be chiral.

The rotations about a C-C single bond that takes a molecule back and forth from trans to gauche can be very fast in a liquid at room temperature. Recent ultrafast infrared laser experiments and theory have determined that the gauche-trans interconversion only takes 50 ps (picoseconds, 10^{-12} s), or 50 trillionths of a second. So in a liquid at room temperature, the two forms of butane will be switching back and forth so fast that it is not possible to isolate them as distinct molecules.

DOUBLE AND TRIPLE CARBON-CARBON BONDS

While it is very easy for rotation to occur about a C-C single bond, this is not true for carbon-carbon double or triple bonds. In Chapter 13, we discussed that O_2 has a double bond and N_2 has a triple bond. Carbon-carbon bonds can be single, double, or triple. Rotation about a C-C double or triple bond is virtually impossible. Therefore, double bonds can lock molecules that are the same structural isomers into different conformers. As we will see in Chapter 16, this is where the term trans fats comes from. However, before we go on to discuss large molecules, such as trans fats, we first need to talk about C-C double and triple bonds.

In the carbon compounds discussed so far, carbon uses four sp^3 hybrid atomic orbitals to make four single σ bonds to other atoms. In such bonding, each carbon has a tetrahedral arrangement of four bonds. In Figure 14.3, formaldehyde is shown. Formaldehyde has a carbon with a double bond. To illustrate the manner in which carbon can make single, double, and triple bonds, we will take a look at the bonding in ethane, ethylene, and acetylene. These three compounds have chemical formulas, H_3C-CH_3, $H_2C=CH_2$, and $HC=CH$, respectively. Ethane has a single bond, ethylene has a double bond, and acetylene has a triple bond. Figure 14.14 shows the structures of the three molecules. In ethane, each carbon forms four tetrahedral bonds. In ethylene, each carbon forms three trigonal bonds, and in acetylene, each carbon forms two linear bonds.

Although in each of the three molecules the two carbons are bonded to each other, the bond order really makes a difference. Table 14.1 gives the bond orders, C-C bond lengths, and bond ener-

FIGURE 14.14. *Ethane, single bond, carbon tetrahedral bonds. Ethylene, double bond, carbon trigonal bonds. Acetylene, triple bond, carbon linear bonds.*

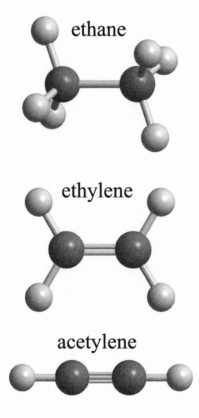

ethane

ethylene

acetylene

TABLE 14.1. *Single, Double, and Triple C-C Bonds.*

	bond order	C-C bond length	C-C energy
ethane	single (1)	1.54 Å	5.8×10^{-19} J
ethylene	double (2)	1.35 Å	8.7×10^{-19} J
acetylene	triple (3)	1.21 Å	16×10^{-19} J

gies of the three molecules. As the bond order increases, the bond length decreases substantially and the bond energy almost triples in going from the single to the triple bond.

Carbon-Carbon Double Bond—Ethylene

First let's look at the bonding in ethylene. As can be seen in Figure 14.15, the carbon centers are trigonal. As discussed, to have trigonal bonding, a carbon atom will use three sp^2 hybrid atomic orbitals to form bonding MOs (see Figure 14.7). Carbon has four valence atomic orbitals to use for bonding, 2s, $2p_x$, $2p_y$, and $2p_z$. In the top part of the figure, the ethylene molecule is in the xy plane. So the carbons and the hydrogens are in the plane of the page, which is xy. To form the trigonal sp^2 hybrids to make three bonds, each carbon will use the 2s, $2p_x$, and $2p_y$ orbitals. With the three sp^2 hybrids, each carbon will make three σ bonds, one to the other carbon and two to hydrogens. The σ bonding is shown in the top portion of Figure 14.15.

When a carbon forms three sp^2 hybrids with the 2s, $2p_x$, and $2p_y$ orbitals, it has its $2p_z$ orbital left over that does no participate in the σ bonding. In the top part of Figure 14.15, the $2p_z$ orbitals stick out of and into the page. Each carbon has one unpaired electron in its $2p_z$ orbital. The bottom portion of the figure shows the ethylene molecule rotated. The σ bonds are shown as the lines connecting the atoms. The positive lobes of the $2p_z$ orbitals overlap construc-

FIGURE 14.15. *Ethylene double bond orbitals. Top: Each carbon uses three sp² hybrids to make three σ bonds in a trigonal configuration. The page is the xy plane, with z out of the plane. Bottom: Each carbon has a $2p_z$ orbital not used to form the sp² hybrids. The $2p_z$ orbitals combine to make a π bonding molecular orbital to give a second bond between the carbons.*

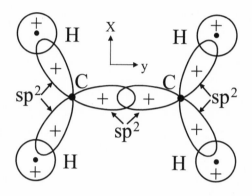

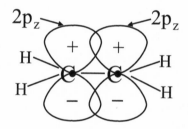

tively as do the negative lobes. The two $2p_z$ orbitals combine to make a π bonding molecular orbital, as shown in Figure 13.3. This is a π bond because there is no electron density along the line connecting the carbon atomic centers. The net result is that the two carbon atoms have a double bond composed of a σ bond formed from a sp² orbital on each carbon and a π bond formed from a $2p_z$ orbital on each orbital.

It is not possible for rotation to occur around the carbon-carbon double bond. Rotation around the bond would require the overlap of the two $2p_z$ orbitals to get worse and worse as the angle got bigger and bigger. For a 90° rotation, the two 2p orbitals would be pointing perpendicular to each other, and have no overlap. Such a rotation would break the π bond, which takes a great deal of energy. As mentioned, measurements and theory have been used to determine that the time for butane in liquid solution to rotate about the C-C single bond is about 50 ps. For ethane the time is about 12 ps. Butane rotates slower than ethane about the single C-C bond because it has two additional methyl groups (CH_3), one on each end of the central two carbons. If you put ethylene in the same liquid at room temperature, it can be roughly estimated that it will take about one hundred billion years to rotate about the double bond because of the large amount of energy required to break the π bond. Therefore, for all practical purposes, a double bond (and a triple bond) prevents rotational isomerization between conformers that differ by the configuration about the double bond.

Carbon-Carbon Triple Bond—Acetylene

Acetylene forms a triple bond between the carbons in a manor analogous to the double bond formation in ethylene. Each carbon has four atomic orbitals, 2s, $2p_x$, $2p_y$, and $2p_z$, to use in bonding. Acetylene is linear (see Figure 14.14). Take the line of the molecule to be the x axis. Then, each carbon will form two sp hybrids from its 2s and $2p_x$ orbitals. The two sp hybrids on a carbon will be used to make two σ bonds, one to the other carbon and one to a hydrogen. That leaves two unused 2p orbitals on each carbon, the $2p_y$ and the $2p_z$. The $2p_z$ orbitals on each carbon form one π bonding MO, and the $2p_y$ orbitals on each carbon form another π bonding MO. The result is that the two carbons have a triple bond, one σ bond and two π bonds.

In the next several chapters, we will discuss a number of important types of molecules, for example, alcohols, organic acids, large hydrocarbons, and carbon containing fuels, that is, coal, oil, and natural gas. Small alcohols are discussed so that we can see what an alcohol is and how small differences in structure really matter if you decide to drink something other than ethanol (the alcohol in beer). The ideas will show why some molecules dissolve in water while others don't, and how soap (a type of large organic molecule) makes insoluble grease dissolve in water. We will examine the importance of the inability of double bonds to undergo rotational structural change in connection with fats and trans fats. What happens when carbon-based fuels are burned will be described, and why, for a given amount of energy produced, one fuel produces more of the greenhouse gas carbon dioxide than another. It is well known that carbon dioxide is a greenhouse gas, but why? We will see how the combination of two fundamental quantum mechanical effects makes carbon dioxide a potent greenhouse gas.

15

Beer and Soap

IN THIS CHAPTER, we will look at several types of molecules to see how differences in their nature influence chemical processes. First, alcohols will be introduced. An alcohol is an organic molecule that contains a particular type of chemical group. An alcohol can be a relatively small molecule, such as ethyl alcohol, which chemists usually call ethanol. Ethanol is the alcohol in beer, wine, and vodka. However, large important biological molecules, such as cholesterol, are also alcohols. We will get to such large molecules in Chapter 16. We will see why ethanol dissolves in water, how it can turn into vinegar, and outline the chemical reactions in your body that makes methanol (wood alcohol) poisonous, but ethanol safe, at least in moderation. Building on the mechanism that enables some molecules to dissolve in water, we will take a look at the structure of soap and oil molecules to see why you need soap to take grease off of your dishes and get it to wash down the drain.

ALCOHOLS

Ethanol is ethane (Figure 14.10) with one of the hydrogens replaced by an OH group. The OH group is called a hydroxyl group. The chemical formula for ethanol is H_3CH_2COH. Figure 15.1 shows a diagram and a ball-and-stick model of ethanol. In ethanol as in ethane, the carbon atoms use four sp^3 hybrid orbitals to form tetrahedral bonds. The oxygen also uses four sp^3 hybrids. One of them is used to make the bond to the carbon, one is used to bond to the hydrogen, and the other two contain electron lone pairs. The lone pairs are not shown in the ball-and-stick model in Figure 15.1. (Figure 14.2 shows the oxygen lone pairs for the water molecule.)

Note that in the ball-and-stick model of ethanol, the hydrogen that is bonded to the oxygen is considerably smaller than the hydro-

FIGURE 15.1. *Ethanol (ethyl alcohol) diagram showing atom connectivity (top), ball-and-stick model (bottom). The hydrogens are light gray, the carbons are dark gray, and the oxygen is black.*

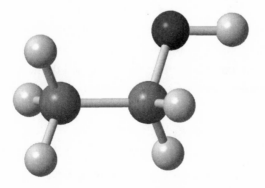

gens bonded to the carbons. Going back to the Periodic Table (Chapter 11), we know that oxygen really wants to pick up electrons to obtain the neon closed shell configuration. It gains the electrons by making covalent electron pair sharing bonds. However, the sharing is not completely equal in a bond between oxygen and hydrogen. The oxygen has a very strong attraction for electrons, so it pulls some extra electron density away from the hydrogen. The extra electron density on the oxygen makes it a little bit negative and the loss of electron density makes the hydrogen a little bit positive. This loss of electron density reduces the size of the electron cloud of hydrogen, which is represented by the reduced size of sphere that represents the hydrogen bonded to the oxygen. Carbon and hydrogen have almost the identical attraction for electrons when they make a covalent bond. They share the electrons almost equally. Thus on average, there is more electron density on a hydrogen bound to a carbon than on a hydrogen bound to an oxygen. In general, an alcohol is a molecule that has a carbon with an OH group and that carbon is also bonded only to hydrogens or other carbons.

ETHANOL IS A LIQUID AT
ROOM TEMPERATURE, NOT A GAS

At room temperature, ethane is a gas but ethanol is a liquid. You have to make ethane colder than $-89°$ C ($-128°$ F) before it becomes a liquid, and you have to make ethanol hotter than $78°$ C ($172°$ F) before it boils and goes from a liquid to a gas. Ethane and larger hydrocarbons like oil do not dissolve in water, whereas ethanol and larger alcohols do dissolve in water. Ethane and ethanol are almost the same size, and they have similar shapes. So why, in contrast to ethane, does ethanol dissolve in water, and why is it a liquid at room temperature?

As we briefly discussed, ethanol's hydroxyl group, the OH, has two partial changes. The O is a little bit negative, and the H is a little bit positive. A diagram showing the partial charges is: $O^{\delta-} - H^{\delta+}$. The δ (Greek letter delta) is used to mean "a little bit." The δ is followed by the sign of the electrical charge on the atom. The amount of electron density transferred from the H to the O is very small, much less than one full electron that is transferred in a salt like NaCl, which is Na^+ and Cl^-. The bond between the oxygen and the hydrogen is mainly covalent, not ionic as in NaCl. However, the partial charges on the O and H are unbelievably important. They result from the details of the quantum mechanical molecular orbitals responsible for the oxygen-hydrogen covalent bond. These partial charges result in ethanol being a liquid. If you will permit only mild hyperbole, without the same type of partial charges on the oxygen and hydrogen atoms of water molecules, life would not exist.

Ethanol is a liquid because the partial changes give rise to a type of chemical interaction between molecules called hydrogen bonds. Hydrogen bonds are much weaker than a real covalent chemical bond. They are about a 10th the strength or less. While an accurate description of hydrogen bond formation requires quantum theory, a qualitative understanding can be obtained by considering electrostatic interactions between partial charges. A hydrogen bond is formed when the slightly positive H on one molecule is attracted to the slightly negative O on a different molecule. That attraction will tend to maintain the H of one ethanol in a pretty well-defined position relative to the O of another ethanol. These attractions hold the ethanols together to form a liquid at room temperature. Ethane molecules do not have these relatively strong intermolecular attractions.

Heat is a form of kinetic energy. Heat jiggles the molecules around. In ethane, the molecules do not have a substantial attraction of one for another. At room temperature, the heat-induced motions cause ethane molecules to fly apart, so ethane is a gas. Imagine you and another person hold hands and run in opposite directions. If you hold hands weakly, you will fly apart and take off in different directions, like ethane molecules. If you hold hands very tightly, the two of you will stay together, but move around somewhat as you tug each other to and fro, like ethanol molecules.

Figure 15.2 shows four ethanol molecules hydrogen bonded together into a chain. The dashed lines go from the hydrogen of the OH group of one ethanol to a lone pair on an oxygen of another ethanol. The lone pair has a good deal of electron density, so the

FIGURE 15.2. *Four ethanol molecules that form a chain. The oxygens are almost black in the figure. An oxygen has two lone pairs in addition to the hydrogen and carbon bonded to it. The dashed lines show hydrogen bonds that go from the hydroxyl's H on one ethanol to an oxygen lone pair on another ethanol.*

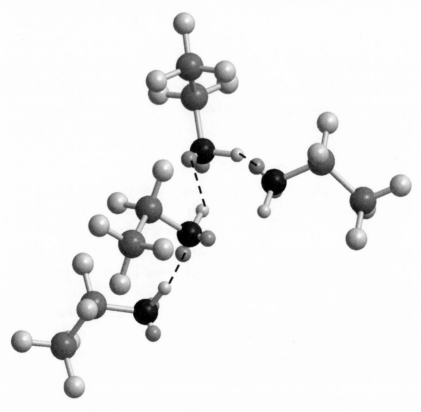

slightly positive H is attracted to the electrons of a lone pair on oxygen. This continues from one ethanol to another to form chains. Ethanol liquid is composed of hydrogen-bonded chains that hold the molecules together. The hydrogen bonds make ethanol a liquid at room temperature, but they are relatively weak. These bonds are constantly being broken and reformed, but on average, each ethanol has hydrogen bonds (H-bonds) to one or more other ethanols. However, if you heat the ethanol up enough, the thermally (heat) induced jiggling causes the molecules to break the H-bonds and fly apart. The temperature at which the thermal energy is enough to cause the ethanols to fly apart is the boiling point (78° C). At this temperature or above, ethanol is a gas.

WATER MAKES HYDROGEN BONDS

Back to why hydrogen bonds are necessary for life. Water, H_2O, is a very small molecule. It has a molecular weight that is similar to that of oxygen (O_2), nitrogen (N_2), or methane (CH_4), all of which are gases at room temperature. Water has a single oxygen bonded to two hydrogens. Like the situation in ethanol, the oxygen makes covalent bonds to the hydrogens, but in an O-H covalent bond, the electrons are not shared perfectly equally. In a water molecule, the oxygen withdraws some electron density from the H's. A diagram of water showing this is: $H^{\delta+} - O^{\delta-} - H^{\delta+}$. The slightly positive hydrogens on one water molecule are attracted to the slightly negative oxygens on another water molecule. One water molecule can make up to four hydrogen bonds.

A schematic illustration of water hydrogen bonding is shown in Figure 15.3. The central water molecule is making four hydrogen bonds with the surrounding four waters. The central water's two hydroxyls are hydrogen bonded to two oxygens of other water molecules. And two other water molecules' hydroxyls are forming hydrogen bonds with the oxygen of the central water. In contrast to the

FIGURE 15.3. *A central water molecule hydrogen bonded to four surrounding water molecules. The central water's two hydroxyl hydrogens
bond to two oxygens, and the central water's oxygen accepts two hydroxyl
bonds from two other water molecules.*

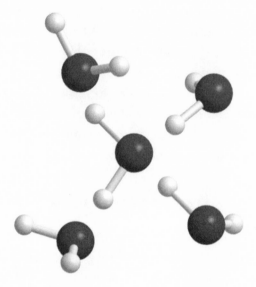

depiction in Figure 15.3, the hydrogen bonding does not stop with
the five water molecules. The outer four water molecules will each
make approximately four hydrogen bonds with other water molecules. The result is a hydrogen bonding network.

At room temperature, there is enough heat so that hydrogen
bonds between water molecules are continually breaking and new
hydrogen bonds are formed with different water molecules. So the
hydrogen bond network is not static. It is continually reforming and
rearranging. The time scale for these hydrogen bond rearrange-
ments has been measured with ultrafast infrared spectroscopy,
and it is approximately 3 ps (picosecond, a trillionth of a second,
10^{-12} s).

Life is based on chemistry that occurs in water. Recent spacecraft sent to Mars are not looking directly for signs of past life, but rather for signs of past liquid water. Liquid water is so fundamental to life that its presence is a necessary and perhaps a sufficient condition for life to exist. Water's amazing properties, which are essential to the biochemistry of life, are a result of its hydrogen bonding network structure and the ability of that structure to rearrange. Water's properties allow it to accommodate a vast array of chemical processes necessary for life. For example, proteins fold in water. Proteins are the very large, extremely complex molecules that are responsible for most of the chemical processes in our bodies. When proteins are chemically produced by other proteins, initially they are not in the correct configuration to function. This is called unfolded. Proteins have regions that will readily form hydrogen bonds to water and regions that are more like hydrocarbon and do not want to mix with water. The protein rearranges its structure, folds, such that the hydrophilic (likes water) regions are on the outside exposed to water and form hydrogen bonds with water and the hydrophobic (dislikes water) regions are on the inside away from water. Such selective interactions with water are an important driving force that helps proteins to obtain the correct shapes necessary to function. It is because water can readily reform its network structure through the making and breaking of hydrogen bonds that it can accommodate the proteins' structural transformation and a vast number of other chemical processes that occur in living organisms.

WATER IS A GREAT SOLVENT

One of water's properties is its ability to dissolve a very wide variety of chemical compounds. We have discussed that NaCl will dissolve in water to give ions, Na^+ and Cl^-. The positive ions are surrounded by the slightly negative oxygens of water, and the negative ions are surrounded by the slightly positive hydrogens of water. Salt dissolves because of water's ability to interact favorably with both cations and anions. Water can also dissolve a very wide variety of organic compounds. Water will not dissolve hydrocarbons such as ethane, but it will dissolve organic molecules like ethanol that have hydroxyl groups (− OH) or other groups that have slightly charged or fully charged portions. Water dissolves ethanol by forming hydrogen bonds to the hydroxyl group of ethanol. In pure ethanol, ethanol molecules hydrogen bond one to another to form chains, as shown in Figure 15.2. When ethanol is put in water, water can form hydrogen bonds to ethanol's hydroxyl, incorporating ethanol molecules into water's hydrogen bonding network. Vodka is essentially ethanol in water. Wine is water with less ethanol in it than vodka. Wine also has large organic molecules that give red wine its color and all wines their distinct aroma and taste.

ETHANOL WILL UNDERGO CHEMICAL
REACTIONS WITH OXYGEN

If wine is exposed to air for too long, it will "go bad" and turn into vinegar. Vinegar is made by purposely making wine go bad. The chemical reactions that convert wine into vinegar are actually facilitated by a bacteria, acetobacter, which has the ability to convert ethanol into acetic acid when oxygen is present. The process requires two chemical reaction steps. The chemical reactions are written as

$$CH_3CH_2OH \rightarrow CH_3CHO + H_2$$
$$2CH_3CHO + O_2 \rightarrow 2CH_3COOH$$

First ethanol (CH_3CH_2OH) is converted to acetaldehyde (CH_3CHO) and hydrogen gas (top line), and then two molecules of acetaldehyde and one oxygen molecule (two oxygen atoms) are converted into two molecules of acetic acid (CH_3COOH), which is vinegar. The structure of ethanol is shown in Figure 15.1. The structures of acetaldehyde (top) and acetic acid (bottom) are shown in

Figure 15.4. In both acetaldehyde and acetic acid, the carbon labeled C_1 forms a methyl group. C_1 is bonded to three hydrogens and carbon C_2. Acetaldehyde has C_2 also bonded to a single hydrogen and double bonded to an oxygen. In general, an aldehyde has a carbon double bonded to an oxygen, bonded to a hydrogen, and bonded to another carbon. In formaldehyde (Figure 14.3), instead of bonding to another carbon, C_2 is bonded to a second hydrogen. C_2 uses three sp^2 hybrid orbitals to form three σ bonds and an additional 2p orbital to combine with a 2p orbital on the oxygen to form a π bond to make the double bond. As shown in the top line

FIGURE 15.4. *Acetaldehyde (top) and acetic acid (bottom). Oxygens are the almost black spheres. Acetaldehyde C_2 carbon is bonded to C_1, a hydrogen, and double bonded to an oxygen. Acetic acid C_2 is bonded to C_1, double bonded to an oxygen and single bonded to another oxygen that is part of a hydroxyl group.*

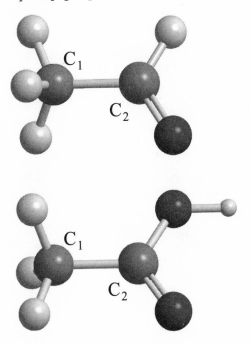

of the chemical reaction equations and looking at the structure of ethanol in Figure 15.1, ethanol goes to acetaldehyde by eliminating two hydrogen atoms to give acetaldehyde and an H_2 molecule. Two molecules of acetaldehyde each pick up one oxygen atom from O_2 to yield two acetic acid molecules (bottom of Figure 15.4). Acetic acid has C_2 bonded to two oxygen atoms. C_2 is double bonded to one and single bonded to the oxygen of the hydroxyl group $-OH$.

Acetic acid is an organic acid, also called a carboxylic acid. The $-COOH$ is the acid group. The simplest definition of an acid is a chemical that when put in water results in the generation of hydrogen ions, H^+. Hydrochloric acid is HCl. Like NaCl, when put in water HCl comes apart to give H^+ and Cl^- ions. HCl dissociates into ions completely. It is called a strong acid because for every molecule of HCl put in water, you get one H^+ ion. The H^+ will be associated with (solvated by) water molecules. The slightly negative oxygens of water molecules surround the H^+. The acid dissociation reaction of acetic acid in water can be written in the following manner.

The $-COOH$ group is the organic acid group. This group ionizes to give $-COO^- + H^+$ as shown. The diagram indicates that the negative charge is stuck on one of the oxygens. In fact, it is shared equally by both oxygens. This equal sharing is represented by the following diagram.

The dashed curve indicates that the molecular orbital that contains the electron that gives rise to the negative charge is spread out over both oxygens. Each oxygen can be thought of as having one-half of a negative charge. Organic acids are very soluble in water. Ethanol is soluble because the hydroxyl oxygen has a small negative charge that leads to hydrogen bond formation with water molecules. Nondissociated acetic acid has two oxygens with partial charges that can hydrogen bond to water. Dissociated acetic acid has two oxygens, each with a half negative charge that readily hydrogen bonds to water.

Organic acids, such as acidic acid, are weak acids. In water, only a small fraction of the acetic acid molecules ionize to produce H^+ and acetate anion. If you have 60 grams of acidic acid in a liter of water, which is about four tablespoons (2 ounces) in a quart of water, only about 0.4% of the acidic acid molecules will be ionized to produce H^+ and acetate anion. This concentration, four tablespoons of acetic acid in a quart of water, is approximately the concentration of acetic acid in common vinegar. It is the acetic acid that gives vinegar its snappy taste. Organic acids are very common in chemistry and biology. All proteins are composed of combinations of 20 amino acids. Prior to reactions to form a protein, each amino acid is a type of organic acid containing the organic acid group − COOH.

METHANOL IS VERY POISONOUS

Methanol is the smallest alcohol molecule. Ethanol is ethane with one hydrogen replaced by a hydroxyl group, − OH. Methanol is methane with one hydrogen replaced by a hydroxyl. While ethanol can be ingested in reasonable quantities without dire consequences, methanol is highly toxic. Methanol, also called wood alcohol, is a common contaminant of moonshine liquor. As little as 20 ml (milliliters) has been known to cause death, and 15 ml can cause blindness. Fifteen ml is one tablespoon. Two tablespoons are about one ounce of liquid. Eight ounces of wine contain about one ounce of ethanol. So replacing the ethanol with methanol in a class of wine can cause blindness and death. This is rather remarkable since ethanol is just methanol with one additional methyl group ($-CH_3$).

It is not methanol itself that is poisonous, but rather the metabolic products of methanol. In humans and other living organisms, alcohols are converted to other chemicals by enzymes (proteins that are responsible for chemical reactions) called alcohol dehydrogenases. In humans, these enzymes are found in the liver and the lining of the stomach. The evolutionary development of these enzymes probably occurred to break down alcohols that are produced by bacteria in the digestive tract or to break down alcohols that are natural components of some foods. Ethanol is converted first to acetaldehyde and then to acetic acid as discussed in connection with Figure 15.4 above, which shows the structures of these two molecules. Acetaldehyde and acetic acid are not harmful to the body and can be readily eliminated. Larger alcohol molecules are also converted first to aldehydes and then to organic acids, which can be eliminated from the body without harm. However, alcohol dehydrogenases convert methanol to formaldehyde and then to formic acid. The structure of formaldehyde is shown in Figure 14.3. Formaldehyde is like acetaldehyde except carbon C_2 of acetaldehyde (Figure 15.4, top) is bonded to a single hydrogen rather than to a methyl group (C_1 in Figure 15.4). Formic acid is like acetic acid (Figure 15.4, bottom) but again with C_2 bonded to a hydrogen rather than a methyl group. Formaldehyde in particular, but also formic acid, are highly toxic. They damage the retina and the optic nerve, leading to vision impairment and blindness. However, formic acid also is responsible for serious acidosis, which involves interfering with enzymes that break down carbohydrates. Relatively low concentrations of formaldehyde and formic acid, as well as other metabolites derived from them, can cause death.

SOAP

As we discussed, alcohols like ethanol and organic acids, such as acetic acid, are very soluble in water because the organic groups containing oxygens can form hydrogen bonds with water. In contrast, ethane, which is very similar to ethanol, does not dissolve in water, because it does not have oxygen containing groups that can form hydrogen bonds. Ethane is a hydrocarbon, that is, it is composed only of hydrogens and carbons. Methane and ethane are gases. Larger hydrocarbons, beginning with pentane (five carbons), are liquids at room temperature. The smaller of these liquid hydrocarbons like pentane and octane (a component of gasoline) are very thin liquids, that is, they are not viscous.

LARGE HYDROCARBONS ARE OIL AND GREASE

As the number of carbons increases, the liquid hydrocarbons become increasingly viscous. The heating oil, used in homes and businesses in many areas of the United States, is composed of a mixture of hydrocarbons typically spanning the range of 14 to 20 carbons. At room temperature, oil flows readily but it is much more viscous than gasoline. Grease is composed of really large hydrocarbons. It is very viscous and will not pour at room temperature.

The hydrocarbons that comprise heating oil are liquids at room temperature, but as discussed, they do not dissolve in water. Molecules with 14 carbons are the lightest component of heating oil. Figure 15.5 shows n-tetradecane. Decane has 10 carbons. Tetradecane has four (tetra) additional carbons. The n (normal) means that all of the carbons are connected one to another with no branches. That is, each carbon is connected to at most two other carbons. The upper portion of the figure shows a ball-and-stick model of n-tetradecane. It is important to remember that the electron density surrounding the atoms in a molecule is space filling. The bottom portion of the figure shows a space-filling model of n-tetradecane.

FIGURE 15.5. *n-tetradecane, $C_{14}H_{30}$, ball-and-stick model (top) and space-filling model (bottom). The molecule has 14 carbons connected one to the next without branching.*

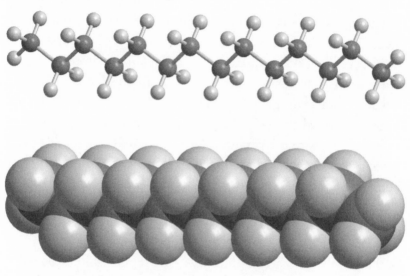

LARGE HYDROCARBONS CAN HAVE MANY STRUCTURES

Many other hydrocarbons have 14 carbons. These are branched. Figure 15.6 shows a ball-and-stick model (top) and a space-filling model (bottom) of one of them, 2,8-dimethyldodecane. Dodecane has 12 carbons. Two additional methyl groups branch off of the main chain at the second and eighth carbon from the left. n-tetradecane and 2,8-dimethyldodecane are structural isomers. They have the same number of hydrogens and carbons, but no amount of rotation about the bonds can convert one into the other. Both n-tetradecane and 2,8-dimethyldodecane have many conformers, that is, it is possible to rotate about various carbon-carbon single bonds to produce different shapes without changing how the carbons are connected. Structural isomers and conformers were discussed in connection with butane (see Figures 14.12 and 14.13).

FIGURE 15.6. *2,8-dimethlydodecane, $C_{14}H_{30}$, ball-and-stick model (top) and space-filling model (bottom). The molecule has 14 carbons. There is a chain of 12 carbons, with two methyl groups branching off at the second and eighth carbons from the left.*

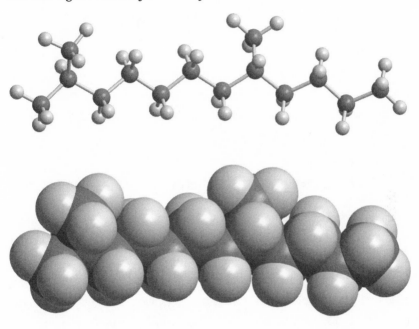

OIL AND WATER DON'T MIX

Heating oil is a relatively viscous liquid, although the hydrocarbon molecules have relatively weak attractive interactions one for another. The large number of sizes, structural isomers, and conformers cause the molecules to become entangled, which contributes to the viscosity. If oil is in water, it will float on top. If you shake it up, it will appear to mix for a while. However, if you let it stand, the oil will separate and again float on top. Anyone who has made their own oil-and-vinegar salad dressing knows this. You mix olive oil, vinegar, and possibly some water, and then you shake it up. If you let it stand, the olive oil floats right back to the top. In commercial oil and vinegar salad dressings, emulsifiers are added to keep the oil and vinegar from separating. Emulsifiers are very similar to the soap that we are about to describe. We know that an oxygen atom in a water molecule is slightly negative and is attracted to atoms that are positively charged or at least have a partial positive charge. The hydrogens of water molecules are slightly positive and are attracted to negatively charged or partially negatively charged atoms. Hydrocarbons have carbons and hydrogens that are essentially neutral in charge. Therefore, water molecules are attracted to each other much more strongly than they are attracted to oil. The result is oil does not dissolve in water.

THE STRUCTURE OF SOAP MOLECULES

Soap makes oil dissolve in water. Many different molecules are used as soaps or detergents. The more formal name for a soap molecule is a surfactant. While the chemical nature and structure of surfactants vary widely, all surfactants have a common feature. A section of a surfactant molecule, if taken by itself, would be very soluble in water. The other part of the surfactant molecule by itself would be very soluble in oil and grease.

One such molecule is sodium n-heptadecaneacetate, which is shown as both a ball-and-stick model and a space-filling model in Figure 15.7. n-heptadecane is an unbranched 17-carbon chain. This hydrocarbon portion of the molecule is shown in the figure as a particular conformer with a couple of rotations about carbon-carbon bonds that produces the bent shape. Tetradecane, shown in Figure 15.5, is all trans. It does not have any rotations to give some gauche conformations. Large hydrocarbons have many different conformers that can interconvert. By itself, heptadecane would be one component of heating oil.

The n-heptadecane hydrocarbon is attached to an acetate group or acetate anion. The acetate group comprises the last two carbons and two oxygens on the right side of the molecule shown in Figure 15.7. The acetate anion is shown on page 260 in the chemical diagram representing the dissociation of acetic acid. For dissociated acetic acid, the cation is H⁺. Here the cation is sodium, Na⁺, which is not shown in Figure 15.7. Sodium acetate is represented in the following diagram.

$$\begin{array}{c} H \\ H\text{-}C\text{-}C \\ H \end{array} \begin{array}{c} O^{\delta-} \\ \\ O_{\delta-} \end{array} + \; Na^+$$

Sodium acetate is a sodium salt like sodium chloride, NaCl. However, here the anion is an organic anion rather than the ele-

FIGURE 15.7. *Sodium heptadecaneacetate, $C_{18}H_{37}COO^-\ Na^+$, ball-and-stick model (top) and space-filling model (bottom). The dissociated sodium ion is not shown. The molecule has 19 carbons. There is a chain of 17 carbons and then an acetate group. The δ^- indicates that each oxygen (darkest spheres) carries an approximately negative one-half charge.*

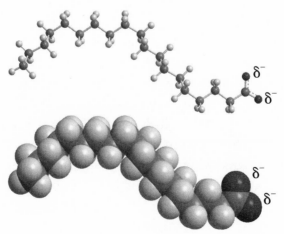

mental anion, Cl⁻. Sodium acetate dissolves completely in water just like NaCl does.

Putting Soap in Water Forms Micelles

So the molecule sodium n-heptadecaneacetate is composed of a long hydrocarbon chain that will not dissolve in water and sodium acetate that readily dissolves in water. What happens if you put a substantial amount of soap, in this case sodium n-heptadecaneacetate, in water with no oil or grease around? The hydrocarbon portions of the molecules hate water, so they want to avoid it. As a pure hydrocarbon, n-heptadecane would completely separate from the water and float on top of it. However, the sodium acetate portion loves water. It will dissociate into an acetate anion and a sodium cation, both of which will have strong favorable interactions with water molecules. Both portions of the surfactant molecule get what they want by forming micelles. Micelles are nanoscopic structures, that is, they are structures with a size on the order of nanometers. A common shape for a micelle is spherical, or near spherical, although a variety of shapes occur, depending on the surfactant and the concentration of surfactant in the water. Their sizes are typically about 10 nanometers, that is, 10 billionths of a meter in diameter. The size and structure of the surfactant molecules determine the size of the micelles.

Figure 15.8 shows a schematic illustration of a spherical micelle. The balls represent the acetate or other charged or hydrophilic portions of the surfactant molecules. The hydrophilic portion of a surfactant molecule is often referred to as the head group. The squiggly lines represent the hydrophobic hydrocarbon tails of the surfactants. The head groups are very soluble in water and form an outer shell. The hydrocarbon tails avoid contact with water by clustering together to form a nanodroplet of oil called the core of the micelle. The formation of micelles enables soap to readily dissolve in water.

FIGURE 15.8. *A schematic of a spherical micelle. The balls represent the acetate anion head groups. Wiggly lines represent the hydrocarbon tails. The micelle is surrounded by water that hydrogen bonds to the head groups. The hydrocarbon chains clump together to form a nanodroplet of oil, which is protected from water by the head groups.*

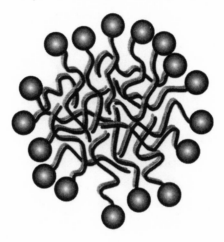

Soap Dissolves Grease

Now consider what happens when plates or hands, with grease or oil on the surface, are put into soapy water. In pure water, the hydrocarbons on a surface are repealed by the water. However, with soap micelles in the water, the situation is very different. The charged head groups of the micelles come in contact with the oily surface. The head groups want to avoid the oil, which causes the micelles to open up, exposing the surfactants' hydrocarbon tails to the grease. The tails of surfactants are perfectly happy to be embedded in the oil and grease. The oily hydrocarbons become entangled with the surfactant tails. Helped by agitation, some of the oil hydrocarbons lift off from the rest of the oily surface. The surfactant head groups close up around the core, reforming a micelle. However, some of the oil and grease hydrocarbons have been incorporated into the micelle's core. The containment of hydrocarbons in the interior of a micelle is shown schematically in Figure 15.9. The hydrocarbon tails of the surfactants are the double lines, while the oil hydrocarbons are single spotted lines. The oil and grease molecules remain in the micelle core as part of the oil nanodroplet. The additional hydrocarbons in the core make the micelles bigger. More surfactant molecules, which are in the water, can join a micelle to fully surround the enlarged oil nanodroplet. The charged head groups of one micelle repel those of other micelles, which prevents the grease from coagulating and forming grease globs that are not soluble in water.

Soap-like materials are reported to have been produced as early as 2800 BCE. True soaps, basically the same as those used today, were made by chemists in the Islamic world in the seventh century. Today, we hear a lot about the coming of nanotechnology, in which nanometer scale assemblies of molecules or atoms can perform

FIGURE 15.9. *A schematic of hydrocarbons from oil or grease (single spotted lines) contained in the interior of a soap micelle.*

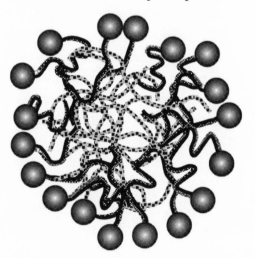

very specialized functions. It is remarkable that soap in water is a nanoscopic material. Surfactants form nanometer-size micelles that can encapsulate grease and oil. The micelles with encapsulated hydrocarbons are soluble in water, which makes it possible for us to wash away otherwise water-insoluble molecules.

16

Fat, It's All About the Double Bonds

IN THIS CHAPTER, we will use some of the ideas developed previously to talk about some of the large molecules commonly discussed in everyday life. We hear about such things as saturated fats, unsaturated fats, transfats, and cholesterol. What are these things, and how do they differ? What are the relationships between molecular structure and health effects?

WHAT IS A FAT MOLECULE?

When we say fat, you may think of butter, lard, olive oil, or cottonseed oil. Each of these is actually a mixture of different fats. Figure 16.1 shows one particular molecule that is a fat. It is stearic acid. It is a long hydrocarbon chain with an organic acid group on the end. Stearic acid has 18 carbon atoms. The end carbon on the right is the organic acid group. Figure 15.4, bottom, shows acetic acid, which is an acid group with one methyl attached. Figure 15.5 shows tetradecane, a 14-carbon hydrocarbon, which is a component of heating oil. Stearic acid is like acetic acid but with a 17-carbon chain attached to

FIGURE 16.1. *Ball-and-stick (top) and space-filling models (bottom) of stearic acid. Stearic acid has 18 carbon atoms, 36 hydrogen atoms, and two oxygen atoms. It is a 17-carbon hydrocarbon with an acid group, — COOH, on the end (right side).*

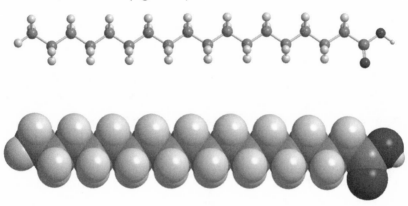

the acid group instead of a single methyl group, or stearic acid is a somewhat longer hydrocarbon than tetradecane with an acid group on the end. Basically, a fat is a long carbon chain with an organic acid group on one end. The acid group forms hydrogen bonds to water. As discussed in Chapter 15, acetic acid is soluble in water because of the strong hydrogen bonding interactions between the acid group and water. Stearic acid, and fats in general, are not soluble in water because of the long hydrocarbon chain. While the acid group is strongly attracted to water (hydrophilic), the long hydrocarbon chain, like hydrocarbons discussed in Chapter 15, does not like to interact with water (hydrophobic). For fats, the long hydrocarbon portion of the molecules wins, and in general fats are not water soluble.

SATURATED AND UNSATURATED FATS

Stearic acid is a saturated fat. Each carbon is bonded to one or two other carbons only by single bonds. There are no double bonds connecting carbon atoms. A saturated fat is a fat with only single bonds between carbon atoms.

Figure 16.2 shows a ball-and-stick model of oleic acid. Like stearic acid, oleic acid has 18 carbon atoms with an acid group on one end. However, it has one double bond between carbons 9 and 10, where the carbon atoms are numbered beginning with the carboxylic acid carbon. Oleic acid is a monounsaturated fat. It is unsaturated because it has a double bond. It is monounsaturated because it has only one double bond. Saturated fats have no double bonds between carbon atoms. In stearic acid, except for the carbon in the acid group, all of the carbons use four sp^3 hybrid orbitals to form bonds. Carbons not at the ends of the stearic acid molecule use two of the four sp^3 hybrid orbitals to form single bonds to the adjacent carbons and two to bond two hydrogens. Each carbon, except for the acid carbon, has a tetrahedral arrangement of bonds to other carbons and hydrogens. Oleic acid's carbons 9 and 10 use three sp^2 hybrids to form three σ bonds, one to a hydrogen and two to the adjacent carbons. Carbons 9 and 10 each use their remaining 2p orbitals, one on each carbon, to form a π bond between them. So carbons 9 and 10 have a double bond, and these carbons have a

FIGURE 16.2. *Ball-and-stick model of oleic acid. Oleic acid has 18 carbon atoms like stearic acid in Figure 16.1, but it has one carbon-carbon double bond between carbons 9 and 10 counting from the acid group.*

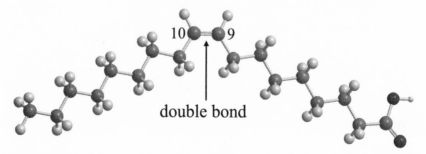

trigonal, rather than a tetrahedral, bonding geometry. The difference in the geometry can be seen clearly in Figure 14.14 by comparing the model of ethane (carbon-carbon single bond) with ethylene (carbon-carbon double bond). In ethane, the carbon centers are tetrahedral. In ethylene, the carbons are trigonal. Oleic acid has 34 hydrogen atoms in contrast to stearic acid, which has 36 hydrogen atoms. Oleic acid uses two orbitals to form the double bond that is used in stearic acid to bond hydrogens. A saturated fat has as many hydrogen atoms as possible, meaning it has no double bonds.

THE SHAPES OF FAT MOLECULES

Stearic acid, as shown in Figure 16.1, is in the all-trans conformation. Figure 14.13 shows butane in the all-trans configuration, with rotation around one of the bounds to give a gauche conformation. Stearic acid can assume many conformations besides the all-trans conformation shown. The all-trans conformation is the most linear, which is the lowest energy conformation for a saturated hydrocarbon or a saturated fat. Because saturated fats only have carbon-carbon single bonds, they are constantly interconverting from one conformer to another. Figure 15.7 displays heptadecaneacetate not in an all-trans configuration. In contrast to stearic acid, oleic acid (Figure 16.2) is not naturally produced in an all-trans configuration. In Figure 16.2, the angles made by carbons 8, 9, and 10 and by carbons 9, 10, and 11 are 120° (trigonal) and not 109.5° (tetrahedral). Therefore, under normal biological conditions adding one double bond locks in a particular shape around the double bond.

SATURATED, MONOUNSATURATED, AND POLYUNSATURATED FATS

Figure 16.3 is a ball-and-stick model of α-linolenic acid. Like stearic acid and oleic acid, α-linolenic acid has 18 carbons, but it has three

FIGURE 16.3. *Ball-and-stick model of α-linolenic acid, which has 18 carbon atoms and three carbon-carbon double bonds.*

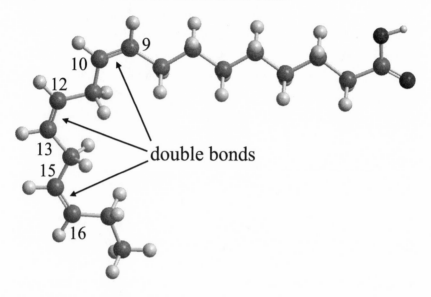

double bonds. α-linolenic acid is said to be polyunsaturated in that it has more than one double bond. In linolenic acid, six of the carbon atoms, specifically 9, 10, 12, 13, 15, and 16, are trigonal, with 120° angles formed by three carbon atoms, such as 8, 9, and 10, rather than with 109.5° tetrahedral angles. Under normal conditions, these additional double bonds force the shape further away from an all-trans configuration. While α-linolenic acid has three double bonds, linolenic acid, which is similar, has only two double bonds. A fat with no double bonds is saturated. A fat with one double bond is monounsaturated, and a fat with two or more double bonds is polyunsaturated.

DOUBLE BONDS IN FATS MATTER

Why does it matter whether or not a fat has double bonds? This is a question that divides into two types of issues, those for chemically modified fats and those for nonchemically modified fats contained in oils and other sources of fat like butter. First, let's look at fats that have not been chemically modified. Fats have been correlated with either increasing or decreasing cholesterol levels. Myristic acid, which is a saturated fat with 14 carbons, is believed to increase cholesterol significantly in a deleterious manner. Palmitic acid, which is a saturated fat with 16 carbons, is thought to increase cholesterol to some extent. In contrast, linolenic acid (18 carbons with two double bonds) and other polyunsaturated fats decrease cholesterol levels. However, some saturated fats, such as stearic acid (Figure 16.1), do not appear to have much influence on cholesterol levels, which is also true of monounsaturated fats such as oleic acid (Figure 16.2).

In common oils used in food, the fractions of important unsaturated and polyunsaturated fats vary widely. Butter fat and coconut oil contain large amounts of myristic acid and palmitic acid and very little linolenic acid. Olive oil contains no myristic acid, but it has a significant amount palmitic acid. It also has some linolenic acid. Canola oil has no myristic acid and almost no palmitic acid. It has a substantial amount of linolenic acid. Grape seed oil, safflower oil, and sunflower oil (the later two when not processed for high-temperature cooking, see below) have no myristic, small amounts of palmitic, but very large amounts of linolenic acid. The chemical composition of these fats indicate that butter fat and coconut oil will have a deleterious effect on blood cholesterol, olive oil is approximately neutral, canola oil has a positive effect, and grape seed oil, safflower oil, and sunflower oil (the last two not for high-temperature cooking) have a very positive effect on blood cholesterol.

CHEMICALLY MODIFIED FATS

Fats are chemically modified for several reasons. Unsaturated fats are fats with double bonds. Double bonds are chemically very reactive. The double bonds in unsaturated fats can react with oxygen. When this occurs to a significant extent, the smell and taste of the oil becomes unpleasant; we say the oil has become rancid. The rates of reactions of oxygen with unsaturated fats are exacerbated by light.

Polyunsaturated fats have more double bonds. With more double bonds available to react with oxygen, the polyunsaturated oil will become rancid more readily. Oils containing unsaturated fats should be refrigerated. Lowering the temperature slows the chemical reactions that cause rancidity. If such oils are not refrigerated, it is better to keep them in a cool, dark place. Some oils come in dark bottles. The dark bottle may improve their shelf life in a store where they are exposed to light. Oils that are composed of almost all saturated fats will keep for a long time without refrigeration. The result is that many oils are chemically processed to reduce or eliminate the double bonds. Such processing also changes the physical properties of the oils, raising the temperatures at which they melt and boil. These changes can be useful in cooking, in preparing baked goods, and in other processed food applications.

PARTIALLY HYDROGENATED AND HYDROGENATED FATS

The oils that are chemically modified to reduce or eliminate double bonds are said to be partially hydrogenated or hydrogenated. A carbon atom makes four bonds. In saturated fats (no double bonds), each carbon is bonded to two other carbons and to two hydrogens, except for the carbons at the ends of a molecule. A double bond eliminates two hydrogen atoms. This can be seen by comparing stearic acid (Figure 16.1) and oleic acid (Figure 16.2). Formation of the double bond in oleic acid between carbons 9 and 10 uses up one bond on each of the two carbon atoms that in stearic acid form bonds to hydrogen atoms. Therefore, the process of eliminating double bonds from a fat increases the number of hydrogen atoms, and we say the oil has been hydrogenated.

HYDROGENATION OF FATS

Double bonds are very stable, and it is difficult to break a double bond. The hydrogenation process turns a carbon-carbon double bond into a carbon-carbon single bond with a hydrogen added to each carbon. The process requires high temperature, a metallic catalyst, and hydrogen. A catalyst is a material that makes a chemical reaction occur faster, but the catalyst is not used up in the process. Qualitatively, the process of hydrogenation works as follows. One of the carbon atoms of a double bond binds to the metal, essentially eliminating the double bond. The binding leaves the other carbon atom with an unpaired electron. That carbon atom is now single bonded to two other carbons and one hydrogen. As we know, a carbon atom wants to make four bonds to have enough electrons to obtain the neon closed shell configuration. The carbon grabs a hydrogen atom. The other carbon atom breaks its bond to the metal catalyst, and it grabs another hydrogen atom. The carbon-carbon double bond has been converted to a single bond and two hydrogen atoms have been added to the fat.

For a polyunsaturated fat, this process can be repeated at all of the double bonds or only at some of them. If it occurs at all of the double bonds, the polyunsaturated fat has been converted to a saturated fat; we say the fat has been hydrogenated. If hydrogenation occurs at some but not all of the double bonds, we say the fat has been partially hydrogenated. The resulting fat may be monounsaturated or still polyunsaturated but with fewer double bonds. The degree of hydrogenation is controlled to produce a resulting fat with the desired properties, particularly whether it is a solid or liquid at room temperature, the melting temperature, and the boiling temperature.

READ THE LABEL

As discussed above, polyunsaturated fats can be beneficial. As they occur naturally, both sunflower oil and safflower oil have very large percentages of polyunsaturated fats. However, many brands of sunflower oil and safflower oil have been partially hydrogenated, so they are more usable for high-temperature cooking. It is possible to tell by reading the nutrition label if sunflower oil and safflower oil have been partially hydrogenated. If the label shows that the amount per serving of monounsaturated fat is greater than the amount of polyunsaturated fat, then the oil has been partially hydrogenated. If these oils have not been partially hydrogenated, then the amount of polyunsaturated fat will greatly exceed the amount of monounsaturated fat. So to get the benefits of significant quantities of polyunsaturated fat from sunflower or safflower oils, the oils should not be partially hydrogenated. Reading the label will tell.

TRANS FATS

All of this hydrogenation of oils sounds fine except for one thing; the big elephant in the room is trans fats. What is a trans fat? Figure 16.4 shows oleic acid in both its cis and trans geometries. Both molecules have 18 carbons, 34 hydrogens, and one carbon-carbon double bond. The atoms are connected one to another in the same order. The difference is the geometry around the double bond.

In the cis conformation, the two hydrogens bonded to carbons 9 and 10 are on the same side of the molecule. They point at an angle toward the top of the page. The molecule is drawn with the double bond horizontal. The angle between the double bond and one of the H atoms is 120° because trigonal sp^2 hybrids are used to make the σ bonds. So the angles made by the bonds from carbons 9 and 10 to the hydrogens are 30° from the vertical direction. For

FIGURE 16.4. *Ball-and-stick models of cis-oleic acid and trans-oleic acid. Both have 18 carbons and one double bond. However, their geometries differ.*

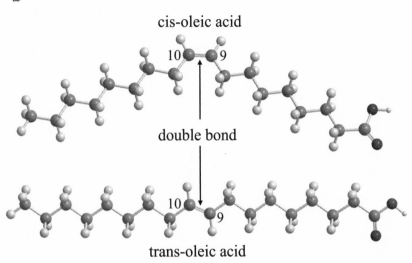

cis-oleic acid

10 9

double bond

10 9

trans-oleic acid

the cis molecule, the two chains of atoms connected to carbons 9 and 10 on either side of the double bond angle downward, again making angles that are 30° from the vertical line that is perpendicular to the double bond.

In the trans conformation, the two hydrogens bonded to carbons 9 and 10 are on opposite sides of the molecule. One points almost straight up toward the top of the page and the other points almost straight down toward the bottom of the page. The two chains of carbon atoms that are bound to carbons 9 and 10 come out in opposite directions relative to the double bond. The net result is that the cis molecule is "bent" around the double bond, while the trans molecule is "straight" relative to the double bond.

Rotation around a carbon-carbon double bond cannot occur under normal conditions. The inability of rotation to occur around double bonds is fundamentally important. Figure 14.13 shows the gauche and trans conformations of n-butane, which has only single bonds. Rotation around single bonds can occur readily at room temperature. Therefore, in the example of n-butane, the gauche and trans conformations are not fixed. In fact, in liquid solution at room temperature, the n-butane gauche and trans conformations will interconvert by rotation around the middle carbon-carbon single bond in approximately 50 ps (50 trillionth of a second), which is a very short time. In contrast, the cis and trans conformations of oleic acid shown in Figure 16.4 are locked in. They will not interconvert without very high temperature and a catalyst.

To understand why rotation around a carbon-carbon single bond occurs readily while rotation about a double bond does not, we need to look again at the hybrid orbitals used by carbon to make single and double carbon-carbon bonds. Figure 14.9 illustrates the hybrid orbitals used in ethane to form the carbon-carbon single bond. Each carbon bonds to the other atoms using four sp^3 hybrid orbitals. The middle part of Figure 14.9 shows schematically the formation of the carbon-carbon single bond by the overlap of one sp^3 orbital on each carbon. The two orbitals, one from each carbon, point right at each other. Rotating one of the carbons does not change the overlap of the orbitals. There are favored configurations because the hydrogens on the two carbons want to avoid each other as much as possible, but the molecule can readily rotate from one favored configuration to another without changing the carbon-carbon sp^3 orbital overlap. This is in contrast to the situation for ethylene, which has a carbon-carbon double bond. Figure 14.15 shows the orbitals used to form the double bond in ethylene. Each carbon uses three sp^2 hybrids to make σ bonds to two hydrogens and the other carbon as shown in the top portion of Figure 14.15. The three sp^2 orbitals on each carbon are formed by superpositions of the 2s, $2p_x$, and $2p_y$ orbitals. These orbitals and the σ bonds are in the plane of the page, which is taken to be the xy plane. That leaves one $2p_z$ orbital on each carbon atom, which will point perpendicular to the plane of the page. As shown in the bottom portion of Figure 14.15, the two $2p_z$ orbitals overlap side to side to form a π bond. If you could grab one of the carbons and rotate it, you would be rotating that $2p_z$ orbital away from the z direction toward the xy plane. Such a rotation will decrease the overlap between the two $2p_z$ orbitals, breaking the π bond. As shown in the table that follows the discussion of Figure 13.9, a double bond is much stronger than a single bond. Therefore, it will take a great deal of energy to rotate around a carbon-carbon double bond because it is necessary to break the π bond for the rotation to occur. The large energy penalty that would be paid prevents rotation.

NATURE PRODUCES CIS FATS BUT CHEMICAL PROCESSING PRODUCES TRANS FATS

Unsaturated fats, both monounsaturated and polyunsaturated, are produced in nature virtually exclusively in cis conformations. Small amounts of trans fats are found in the meat and milk of cattle, sheep, goats, and other ruminants. However, large amounts of trans fats are present in partially hydrogenated oils, and trans fats are also found in hydrogenated oils because the chemical processing does not result in an oil that is 100% saturated. Unprocessed monounsaturated and polyunsaturated vegetable oils have only cis conformations about their double bonds. Partial hydrogenation of the naturally occurring oils generates large quantities of trans fats. This transformation from cis to trans occurs during the hydrogenation process. As discussed above, one of the carbons of a carbon-carbon double bond binds to a metal catalyst in the reaction vessel, which is at very high temperature. When bound to the catalyst, the carbon-carbon bond is effectively a single bond, and rotation from cis to trans can occur. The oil can break its bond to the catalyst before hydrogenation occurs, so the double bond is not hydrogenated, and it reforms. So, rotation from cis to trans can take place before the oil is released from the catalyst. If this occurs, the result is the conversion of a cis conformation to a trans conformation without hydrogenation of the double bond. The processing is intended to reduce the number of double bonds, not eliminate all of them. But a substantial number of double bonds are converted from cis to trans. The result is that partially hydrogenated oils can contain substantial fractions of trans double bonds.

TRANS FATS CAN BE DELETERIOUS

Trans fats have been shown to have a number of deleterious health effects. The basic reason for the harmful effects of trans fats arises from the fact that biological systems have developed to deal with cis fats, and shape matters. Enzymes are proteins (large biological molecules) that act as very specific chemical factories. They can convert fats into other useful molecules, as well as break down fats for elimination. However, an enzyme that will work on a cis fat will, in general, not produce the same chemical reactions or any reaction at all for the trans fat, although it has the identical chemical formula. So two fats that have the identical numbers of carbons, hydrogens, and oxygens, all connected to each other in the same order, will be treated very differently biochemically depending on whether they are cis or trans. Our bodies have not evolved to deal with substantial amounts of trans fats.

Trans fats have been strongly linked to heart disease because of their effect on cholesterol levels in the blood. Trans fats may also have a harmful effect on the nervous system. Myelin is a covering that protects neurons. It is composed of about 30% protein and about 70% fat. The two main fats are oleic acid (see Figures 16.2 and 16.4) and docosahexaenoic acid (DHA, see below). Trans fats replace DHA in brain cell membranes and in myelin. Trans fat alters the electrical signals that comprise the messages in the nervous system, affecting communication between neurons. It is remarkable that a change in shape of a molecule, without changing its chemical composition, can make a beneficial food a harmful food.

WHEN ZERO IS ZERO

Because of the mounting evidence that trans fats are harmful to our health, they should be avoided. Consumer advocacy groups have encouraged the required removal of trans fats from cooking oils used in fast food restaurants and in various commercial products. Because of the bad publicity surrounding trans fats, manufacturers try to avoid alerting consumers to their presence. Now that most people know that partial hydrogenation will produce trans fats, some food labels use the term modified instead of the phrase "partially hydrogenated." Even more amazing is the U.S. government's definition of 0% trans fat. Government regulations permit a product label to say that an oil contains 0% trans fat if one serving contains less than 0.5 grams of trans fat, but the manufacturer is permitted to define the size of a serving. Let's say that there are 0.6 grams of trans fat in a tablespoon of oil. One tablespoon is three teaspoons. So, the manufacturer defines a serving as two teaspoons, which contain 0.4 grams of trans fat. Thus, by changing the definition of a serving, the oil has 0% trans fat. This type of labeling is not permitted in Europe and other countries. Minimizing the amount of partially hydrogenated oil that you consume will reduce your exposure to trans fats.

OMEGA 3 FATTY ACIDS

Figure 16.5 is a ball-and-stick model of docosahexaenoic acid (DHA). As mentioned, DHA is an important component of the lining of nerves. It has 22 carbons and six carbon-carbon double bonds. It is highly unsaturated. All of the double bonds are in a cis

FIGURE 16.5. *Ball-and-stick model of docosahexaenoic acid (DHA). DHA is a polyunsaturated fat with 22 carbon atoms and six carbon-carbon double bonds all in the cis conformation.*

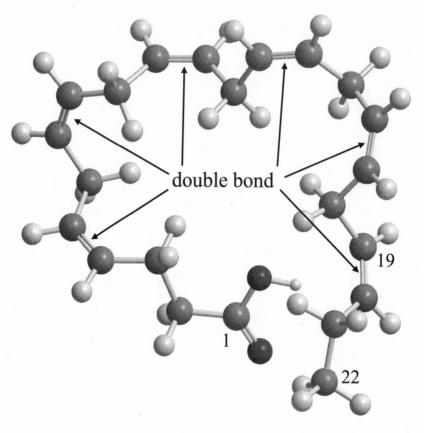

configuration. DHA is one of a class of unsaturated fatty acids (fats) popularly called omega-3 (ω-3) fatty acids. These fats are thought to be beneficial to human health.

The better name for this class of fats is n − 3 fatty acids, with n referring to the number of carbons. The carbons are numbered beginning with the carbon that forms the carboxylic acid. This carbon is labeled 1. So, for DHA counting around the chain, the last carbon at the opposite end from the carboxylic acid group is carbon

22 (see Figure 16.5). This is n, the number of carbons in the chain. Carbon n − 3 is the carbon number that results when 3 is subtracted from the total chain length, n. For DHA, that number is 19 as indicated in Figure 16.5. The fat is an ω-3 fatty acid if the n − 3 carbon is double bonded as shown in the figure. α-linolenic acid displayed in Figure 16.3 is also an ω-3 fatty acid. It has 18 carbons, so n − 3 is 15. As seen in Figure 16.3, there is a double bond between carbons 15 and 16.

TRIGLYCERIDES

The fats we have discussed so far involve single chains. However, a large class of fats commonly found in the body contains three fatty acid molecules tied together into one molecule. These are called triglycerides. Capric acid is a saturated fat with 10 carbons. Capric acid triglyceride is shown in Figure 16.6. There are three capric acid chains, with the first carbon in each chain labeled 1 and the last carbon labeled 10. There is a short carbon chain of three carbons labeled A, B, and C. In a stand-alone capric acid, the acid carbon (labeled 1) is doubled bonded to one oxygen and single bonded to a hydroxyl group, − OH. In a triglyceride, the H of the hydroxyl group is replaced with one of the carbons in the three-carbon chain. In Figure 16.6, the top chain is bonded to carbon A, the middle chain to carbon B, and the bottom chain to carbon C. Capric acid triglyceride is a medium chain-length triglyceride. The medium chain-length triglycerides, or MCTs, have chains from 6 to 12 carbons. Long-chain triglycerides (LCTs) have chain lengths greater than 12.

CHOLESTEROL

When talking about foods and fat, you frequently hear that it is unwise to eat too much fat because it will increase your cholesterol

FIGURE 16.6. *Ball-and-stick model of capric acid triglyceride, which is composed of three capric acid chains. Each chain is a saturated fatty acid with 10 carbon atoms labeled 1 to 10. These are attached to three carbons labeled A, B, and C.*

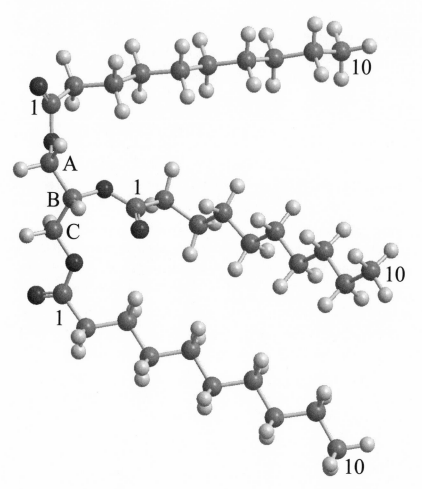

level. Therefore, many people have the mistaken impression that cholesterol is a fat. They think that eating a lot of fat means eating a lot of cholesterol. However, cholesterol is not a fat. Rather, it is an alcohol, as indicated by the suffix, ol. The ol indicates that a molecule is an alcohol, as for example, in ethanol (see Figure 15.1). The structure of cholesterol is shown in Figure 16.7. The top is a diagram of cholesterol, the middle portion is a ball-and-stick model, and the bottom shows a space-filling model. The alcohol − OH group is on the left side of the diagram and is the bottom left group in the ball-and-stick and space-filling models. The molecule has four carbon rings labeled 1 to 4. In the figure, carbons are at all of the vertices, and each carbon will make four bounds. Hydrogens are not shown in the figure except where necessary to indicate if a hydrogen is pointing into or out of the plane of the page. A dashed triangle is pointing into the plane, and a solid triangle is pointing out of the plane. If there is no H at the end of the triangle, then the group is a methyl, − CH_3. The top portion of the figure makes it easy to see which atoms are connected to each other. The ball-and-stick model provides a more detailed three-dimensional illustration of the molecular structure. The space-filling model gives a more representative picture of the molecule's three-dimensional structure. The space-filling model reflects the regions of space where most of the electron probability is located. It is important to keep in mind that molecules are not balls and sticks but rather delocalized electron clouds surrounding the atomic centers, the positively charged atomic nuclei.

Comparing the structure of cholesterol in Figure 16.7 to any of the models of fats (fatty acids) displayed above shows that cholesterol is a very different type of molecule from fats. For example, the space-filling model of stearic acid (Figure 16.1) is very different from the space-filling model of cholesterol in Figure 16.7. Clearly, at the molecular level, cholesterol has little relationship to fatty

FIGURE 16.7. *Cholesterol. Top: Diagram of cholesterol. Middle: Ball-and-stick model. Bottom: Space-filling model. Cholesterol is an alcohol (−OH group) composed of four carbon rings, labeled 1 to 4, and a hydrocarbon chain.*

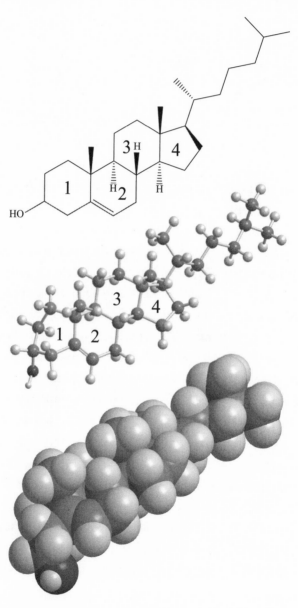

acids. Yet it is frequently linked to discussions of fat in food, and the cholesterol molecule has a very negative aura associated with it.

Cholesterol Is Good, Contrary to Public Perception

Well, cholesterol is getting a bad rap. Cholesterol is a fundamentally important biological molecule. Cells are surrounded by membranes. Inside the cell are all of the complex molecular machines necessary to perform the chemistry responsible for life. Outside the cell are a large number of other chemicals, including oxygen, salts, and large biological molecules. The cell membrane that separates the inside from the outside permits certain molecules to go in and out of the cell while others remain only on the outside or inside. The principal components of these cell membranes are phospholipids. A phospholipid is composed of two hydrocarbon chains, typically 16 carbons long, connected at one end by a head group structure that contains both a positive and a negative charge. The charges of the head group make the head groups highly hydrophilic (attracted to water). The hydrocarbon chains are very hydrophobic (repelled by water). Cells are surrounded by water and also contain a great deal of water on the inside. The charged head groups want to be in water, while the hydrocarbon tails want to avoid water. To satisfy both the hydrophilic-charged head groups and the hydrophobic hydrocarbon tails, the phospholipids arrange themselves into a bilayer, as shown schematically in Figure 16.8.

The figure shows a section of a phospholipid bilayer membrane that completely surrounds and encloses cells. The balls are the charged head groups, and the wavy lines represent the hydrocarbon chains. An actual cell membrane is much more complex than illustrated in Figure 16.8. The membrane contains a large number of proteins that perform specific functions, such as allowing certain ions or molecules to pass into the cell while preventing others from entering.

FIGURE 16.8. *Schematic of a portion of a phospholipid bilayer with two cholesterol molecules. The head groups (balls) are charged and want to be in water. The hydrocarbon tails avoid water by formation of the bilayer. The cholesterol hydroxyl is at the water interface.*

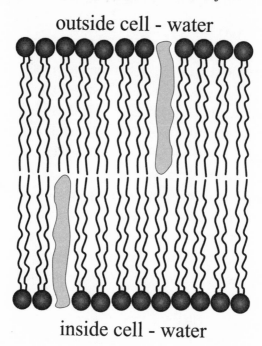

In addition to phospholipids, a major component of cell membranes is cholesterol. Cholesterol comprises as much as 30% of cell membranes. Figure 16.8 shows a schematic of two cholesterol molecules replacing two of the phospholipids. Cholesterol is important because it controls the mechanical properties of the bilayers. Without cholesterol, cell membranes would not function. Therefore, cholesterol is essential. The human body produces a great deal of cholesterol, and only a limited fraction of the necessary cholesterol is taken in through food. The take-home message is that if you could eliminate all of the cholesterol from your body, you would die.

The Problem with Cholesterol

The problem with cholesterol is not that you take some in when you eat, but rather how it behaves in the body. Harmful effects on health from cholesterol are related to fat, but not because cholesterol is a fat or even because fatty foods may contain cholesterol. Cholesterol moves through the bloodstream associated with very large biomolecular aggregates called lipoproteins. These are composed of very large proteins, phospholipids, fats, cholesterol, and other molecules. The lipoproteins can be divided into at least two classes: low-density lipoproteins (LDL) and high-density lipoproteins (HDL). They are somewhat egg shaped, with an approximate diameter of 200 Å (200 $\times$ 10^{-10} meters). The volume of these particles is approximately 5,000,000 Å^3. In contrast, the volume of a cholesterol molecule is approximately 200 Å^3. So, an LDL or HDL particle is about 20,000 times larger than a cholesterol molecule and carries many cholesterol molecules in the bloodstream. High levels of LDL relative to HDL are strongly correlated with coronary artery disease and atherosclerosis. The mechanism is not well understood, but LDL-carrying cholesterol leads to deleterious deposits in the arteries, while HDL does not. These high levels of LDL, when compared to HDL (a high LDL to HDL ratio), are produced by saturated fats and, even worse, trans fats. Saturated fats increase the level of LDL. Trans fats not only increase LDL levels, but they also decrease HDL levels, exacerbating the problem. So eating fat does matter, but not because fatty foods can contain cholesterol. What matters is the nature of the fat we eat. Oils that are high in polyunsaturated fats, which have not been processed in a manner that produces large quantities of trans fats, are desirable.

In Chapter 14, we discussed carbon-carbon single and double bonds. The types of hybrid atomic orbitals used to form molecular orbitals were explicated. Quantum theory allows us to understand the details of bonding and the effects of the nature of bonding on the shapes of molecules and the strengths of the bonds that hold atoms together. In this chapter, fats have been used to illustrate how the seemingly small details of the molecular bonding, such as single bonds versus double bonds, the number of double bonds, and cis versus trans structure about a double bond, play fundamentally important roles in biology. The geometry of double bonds may literally be a life-and-death issue.

17

Greenhouse Gases

IN THIS CHAPTER, we will look at what happens when we burn coal, oil, or natural gas in power plants to create energy. A major point is why coal produces so much more of the greenhouse gas, carbon dioxide, than does oil, which, in turn, produces more than natural gas per unit of energy. Furthermore, the reason that carbon dioxide is such an important greenhouse gas is caused by the fundamental quantum mechanical phenomena of black body radiation and quantized energy levels.

CARBON DIOXIDE FROM
BURNING FOSSIL FUELS

Chapter 15 discussed turning wine (ethanol) into vinegar (acetic acid) by adding oxygen to ethanol. When this happens, we say that ethanol has been oxidized to acetic acid. Oxidation is a chemical process that can take many forms, but in the case of turning ethanol into acetic acid, it literally involves the addition of oxygen. The process is facilitated by biological enzymes. Hydrocarbons, such as methane or heating oil, can also be oxidized. However, hydrocarbons are very stable molecules. They will only oxidize at high temperatures. Burning fossil fuels is the process of oxidization. It takes heat to get the process going, but once the oxidation starts, the breaking of chemical bonds and the formation of new molecules liberates additional heat (thermal energy) that keeps the process going.

BURNING METHANE: NATURAL GAS

First, consider what happens when we burn methane (natural gas). A model of methane is shown in Figure 14.1. Methane (CH_4) reacts with oxygen to give water (H_2O) and carbon dioxide (CO_2). The reaction can be written as follows.

$$CH_4 + 2O_2 \rightarrow 2H_2O + CO_2$$

This chemical equation shows that one molecule of methane will react with two molecules of oxygen to give two molecules of water and one molecule of carbon dioxide. The arrow points from the reactants to the products. This reaction is said to be balanced because the number of carbon, hydrogen, and oxygen atoms is the same on both the left and right sides. In a chemical reaction, the combination of atoms that make up molecules changes, but the number of each type of atom never changes. Actually, another reaction product is heat. It takes energy to break the $C-H$ bonds of methane. However, energy is released when the $O-H$ and $C-O$ bonds are formed to make the products. More usable energy (called free energy) is released in making the bonds to form water and carbon dioxide than is used to break the methane bonds. The net result is that burning methane releases energy that can be used to do things such as boil water to cook spaghetti or drive a steam turbine to make electricity.

WHAT IS A GREENHOUSE GAS?

Methane is a very good fuel, but it produces the greenhouse gas, CO_2. What is a greenhouse gas? A real greenhouse, in which flowers or tomatoes are grown, is a building that lets in a lot of sunlight. Today, such buildings may be constructed with large expanses of plastic that are transparent to sunlight. So the sunlight pours in. When the sunlight lands on the materials inside the greenhouse, much of it is absorbed and converted to heat. You probably have experienced this effect if you have gotten into a car with black or dark color seats that have been illuminated by the sun through the windshield. The seats will be very hot. A black steering wheel may, in fact, even be too hot to touch. As discussed in connection with Figure 9.1, hot objects give off black body radiation, which is light with a wide range of colors. The hotter the object, the higher the frequency of light. The sun is very hot and gives off a great deal of visible light (see Figure 9.1). A black car seat heated by the sun is not very hot and gives off low-frequency (long wavelength) black body radiation. This long wavelength light is in the infrared portion of the electromagnetic spectrum. It is much lower in energy than visible light. In a greenhouse, the sunlight heats up the interior, but the energy that is emitted as infrared black body radiation cannot pass through plastic or glass. These materials are transparent for visible light, but not infrared light. Thus, the heat from the sun is trapped inside the greenhouse, which remains warm even if the outside temperature is much colder.

Carbon dioxide (water vapor and some other gases) makes the atmosphere act as a greenhouse for our entire planet. A great deal of energy falls on the Earth in the form of sunlight. The sunlight heats the Earth's surface, and some of the heat energy is reradiated as infrared black body radiation. The atmosphere is predominantly composed of gaseous oxygen (O_2) and nitrogen (N_2). These gases are transparent in both the visible and infrared parts of the spectrum. If the atmosphere was composed only of oxygen and nitrogen, all of the black body radiation emitted by the warm surface of the Earth would fly right back out into space. The Earth would be much colder than it is and probably not suitable for human life. However, the atmosphere contains other gases. It is approximately 78% nitrogen, 21% oxygen, 0.9% argon, and 0.038% carbon dioxide. In addition, there are traces of other gases and a variable amount of water vapor in the air. The concentration of CO_2 in the air is very small but very important. CO_2 is transparent to visible light, but it absorbs infrared light. (The reasons that carbon dioxide absorbs important wavelengths of infrared light, making it such a serious greenhouse gas, are described below.) So CO_2 lets the sunlight fall on the Earth's surface, but it absorbs some of the infrared black body radiation, preventing it from escaping into space.

A good portion of the infrared black body radiation does escape into space. However, the balance is very delicate. Sunlight is heating the Earth. Black body infrared radiation emitted into space is cooling the Earth. Absorption of infrared by CO_2 in the air reduces the cooling effect. With too little CO_2 in the air, too much heat energy is radiated into space, and the Earth is too cold. But with too much CO_2 in the air, not enough heat energy is radiated into space, and the Earth is too hot. The CO_2 acts like the glass or plastic windows in a real greenhouse. It traps heat on the inside, in this case, the inside of the atmosphere.

Today the concentration of CO_2 in the air is 0.038%, or 380 ppm (parts per million by volume). In 2000, it was 368 ppm. In 1990, it was 354 ppm. In 1980, it was 336 ppm. In 1970 it was 325 ppm. In 1960 it was 316 ppm. These numbers are from measurements made at Mauna Loa, Hawaii. From air trapped in ice, it was found that in 1832 the CO_2 concentration in the air was 284 ppm. There is a clear trend in the CO_2 concentration, and a great deal of scientific work has unmistakably demonstrated that the increase in atmospheric CO_2 is produced by human activity. A major contributor to the increase in CO_2 in the air is the burning of fossil fuels, although other activities, such as clear cutting of rain forests, are also contributors. What happens if the CO_2 concentration in the atmosphere continues to rise? Venus is an extreme example. Its atmosphere is more than 90% CO_2, and its temperature is approximately 900° F (480° C).

BURNING FOSSIL FUELS PRODUCES CARBON DIOXIDE

As we saw in the chemical equation for burning methane (reacting methane with oxygen), the chemical reaction produces CO_2. Burning other fossil fuels also produces CO_2. As discussed, heating oil is a mixture of long chain hydrocarbons, ranging from 14 to 20 carbons. We will use the 14-carbon molecule, tetradecane, as an example. Tetradecane is $C_{14}H_{30}$. The chemical equation for burning tetradecane is

$$C_{14}H_{30} + 21.5O_2 \rightarrow 14CO_2 + 15H_2O.$$

Tetradecane has 30 H atoms, which make 15 water molecules, each with two H atoms. It also has 14 C atoms, which go into the 14 carbon dioxide molecules. To make the $14CO_2$ and $15H_2O$ molecules requires 43 oxygen atoms, or 21.5 O_2 molecules. That is the reason for the $21.5O_2$ on the left side of the chemical equation. Notice that in the chemical equation for burning methane, twice as many water molecules than carbon dioxide molecules are produced. In burning tetradecane, there are approximately the same number of water and carbon dioxide molecules produced. This will turn out to be important.

In addition to natural gas (methane) and oil (long chain hydrocarbons), the third common fossil fuel is coal. In its idealized form, coal is pure carbon. This is not actually true, but for right now we will accept this statement. Then, the chemical equation for burning coal is

$$C + O_2 \rightarrow CO_2.$$

So, in contrast to burning hydrocarbons, burning coal does not produce any water, just CO_2. In burning hydrocarbons, carbon-carbon and carbon-hydrogen bonds must be broken, which costs energy. Then, carbon-oxygen and oxygen-hydrogen bonds are formed to make carbon dioxide and water, which produces energy. Coal also has bonds that must be broken. These are carbon-carbon bonds. Initially, we will take coal to be graphite, which is pure carbon. We want to compare the amount of energy produced by burning each type of fossil fuel with how much of the greenhouse gas, CO_2, is produced. Although graphite is not used as a fuel because it is difficult to ignite, its well-defined chemical structure makes it a useful example.

Energy Production and the Amount of Carbon Dioxide

First we will look at an idealized picture of energy production from burning fossil fuels. We ignore the fact that fuels are not pure and that a good deal of energy is lost in power plants. The actual energy production from real fuels will be discussed below. The three chemical equations for burning the fossil fuels are:

$$CH_4 + 2O_2 \rightarrow 2H_2O + CO_2$$
$$C_{14}H_{30} + 21.5O_2 \rightarrow 15H_2O + 14CO_2$$
$$C + O_2 \rightarrow CO_2.$$

In the first and third equations for burning natural gas and coal, a single CO_2 is produced in the reaction. For our model of heating oil (tetradecane), 14 CO_2s are produced. We want to determine the amount of energy produced for one CO_2. Using thermodynamics, it is possible to calculate the maximum usable energy (free energy) produced by each reaction. We assume that all of the reactions start at room temperature with methane (a gas), tetradecane (a liquid), and graphite (a solid). Of course, burning fuel will initially leave the products hot, but we will consider the situation after everything has cooled to room temperature. For natural gas, we just use the free energy produced by burning one molecule; for graphite, we use the energy produced by burning one carbon atom. For tetradecane, we divide the energy produced by burning one tetradecane by 14 to get the energy per one carbon dioxide molecule produced.

The results are:

methane (natural gas) 1.4×10^{-18} J free energy generated per CO_2 produced

tetradecane (heating oil) 1.1×10^{-18} J free energy generated per CO_2 produced

graphite (coal) 0.7×10^{-18} J free energy generated per CO_2 produced

We see that getting the same amount of energy from coal generates twice as much carbon dioxide (greenhouse gas) as burning natural gas. Coal is also a factor of 1.6 worse in terms of greenhouse gas production for the same amount of energy produced than heating oil. Heating oil is a factor of 1.3 worse than natural gas.

Burning Real Fossil Fuels

The numbers given here are accurate except for coal. The different types of coal, that is, anthracite, bituminous, subbituminous, and lignite (brown coal), produce different amounts of energy per pound and also have different average carbon contents. Even for the same type of coal, the energy content and the carbon content vary. For example, bituminous coal is the most plentiful type in the United States. Its carbon content ranges from 45% to 86%, and its energy content varies by approximately ±20% around the average. The calculated value for coal given above using graphite as a model produces a result corresponding to the middle of the range of energy content of bituminous coal. Natural gas can have as much as

20% of the gaseous hydrocarbons ethane (C_2H_6), propane (C_3H_8), and butane (C_4H_{10}), in addition to methane (CH_4). The mixture hardly changes the energy content or the energy per CO_2 produced compared to that calculated for pure methane. The same is true of oil, which is a mixture of long chain liquid hydrocarbons.

Real Numbers for Carbon Dioxide Production from Generating Electricity

The energy content of fossil fuels does not take into account power plant efficiencies, that is, the conversion of the fuel's energy into electricity. The efficiencies of power plants, which vary depending on design and age, are similar regardless of the type of fossil fuel used. Typically, the efficiency is in the high 30s to 40% range. This means that only approximately 40% of the fuel's energy is converted to electrical energy. In addition, the loss in power transmission lines is approximately 7%. Thus, if a plant efficiency value of 38% is combined with the transmission line loss the overall rate of conversion of fossil fuel energy to usable electricity in our homes is approximately 35%.

To understand what this means, let's calculate how much CO_2 is produced by a 100-watt lightbulb that burns 24 hours a day for a year. A watt is a J/s (one joule for one second). A year is 3.2×10^7 seconds. So, our 100-watt lightbulb will consume 3.2×10^9 J in a year. For natural gas, we said that one molecule of CO_2 is produced for each 1.4×10^{-18} J of chemical energy produced. To get 3.2×10^9 J of energy, we will produce $3.2 \times 10^9/1.4 \times 10^{-18} = 2.3 \times 10^{27}$ molecules of CO_2. This amount is for perfect efficiency. With 35% overall efficiency, 6.4×10^{27} molecules of CO_2 will be produced. 6.02×10^{23} molecules of CO_2 weighs 44 g (grams). So, the weight in grams of the CO_2 produced is 5×10^5 g or 1000 pounds. What we see is that running a 100-watt lightbulb for a year will put 1000 lb of CO_2 into the atmosphere using natural gas. If we burn coal, it will be 2000 lb, or one ton. That is the weight of a small car. The first message is turn off the lights when you aren't using them. That computer you leave on 24/7 uses 200 to 300 watts of electricity. So, if you get your electricity from a coal-fired power plant, your computer is responsible for putting 2 to 3 tons of CO_2 into the atmosphere every year. The second message is that efficiency of the electrical devices we use and the choice of fossil fuel really matters. A compact fluorescent bulb that produces the same amount of light as a 100 W conventional bulb uses only 25 W. So a compact fluorescent running for a year from a natural gas power plant produces 250 lb of CO_2, which should be compared to a conventional bulb run from a coal power plant that produces 2000 lb of CO_2.

CARBON DIOXIDE IS A GREENHOUSE GAS BECAUSE OF QUANTUM MECHANICS

Why is carbon dioxide such a serious problem as a greenhouse gas? In other words, why does it trap heat in the atmosphere? And, why is water vapor (gas phase water molecules in the atmosphere) actually worse than CO_2 as a greenhouse gas? The amount of water vapor is determined by evaporation and condensation of water. The Earth has huge pools of water, the oceans, from which water evaporates into the atmosphere. We also have rain, dew, and snow that remove water from the air. Humans have little effect on the amount of water in the air, although if the Earth continues to warm, the atmosphere will have more water vapor. The additional water vapor will amplify the influence of adding CO_2, making greenhouse warming worse. H_2O is a powerful greenhouse gas. However, we can influence the amount of CO_2 in the air by the energy sources we use and the efficiency with which we use them. CO_2 (and water) are serious greenhouse gases for reasons that come right out of quantum theory.

THE EARTH'S BLACK BODY SPECTRUM

In Chapters 4 and 9, we discussed black body radiation. Figure 9.1 shows the black body spectrum of the sun, whose surface temperature is almost 6000° C. The black body emission has a good deal of its intensity in the visible portion of the spectrum, with a substantial amount of light being emitted in the ultraviolet and the near infrared portions. The colors emitted by a hot object depend on the object's temperature. Hotter objects emit shorter wavelengths. The Earth is, of course, much colder than the sun. Nonetheless, it is a black body emitter. Still, the wavelengths it emits are very long (very low energy photons). Sunlight with the spectrum given in Figure 9.1 falls on the Earth. Some of this light is reflected back into space by ice and other highly reflecting surface features. However, much of the light energy is converted into heat, which warms the Earth. Black body emission by the Earth radiates some of the energy that comes from the sun back into space.

The top portion of Figure 17.1 shows three calculated black body spectra of the Earth for three temperatures. The three curves are normalized, that is, their amplitudes adjusted to all have a maximum value of 1. 15° C (59° F) is the average surface temperature of the Earth; 27° C (81° F) is the surface temperature in the tropics; and $-16°$ C (3° F) is the surface temperature in subarctic regions. While the curves vary somewhat, they are, by and large, very similar. The differences do not matter for discussing the influence of carbon dioxide.

Absorption of the Earth's Black Body Radiation

The bottom two spectra in Figure 17.1 (note the frequency scale is different from the top spectrum) show the atmospheric transmission of infrared radiation through carbon dioxide and water in the long wavelength regions. A transmission of 1 means all of the light passes through the atmosphere. Zero transmission means that

FIGURE 17.1. *Top: Calculated Earth black body spectra for three temperatures (solid curves). The shaded regions show the portions of the spectrum that are strongly absorbed by water and carbon dioxide in the atmosphere. Middle and bottom: spectra of the strong absorption by carbon dioxide and water in the range of 0 to 1000 cm^{-1}. Note the scale difference with the top part of the figure.*

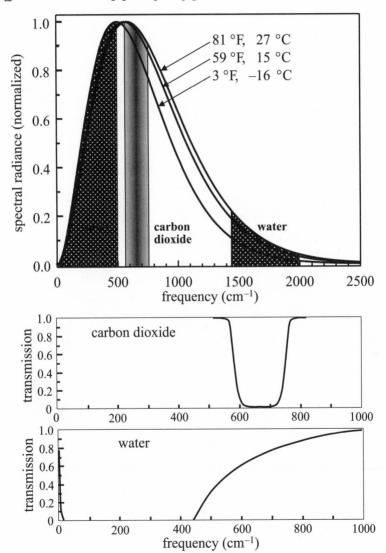

none of the light passes through the atmosphere. These spectra differ depending on which region of the Earth they are measured. The spectra shown are representative. In addition, a good deal of fine structure (peaks and troughs), particularly in the water spectrum, is not shown. The purpose of these curves is to display the essential features of the infrared absorption by carbon dioxide and water that are in the intense part of the Earth's black body spectrum. These absorptions are indicated by the shaded regions. Water also has a strong absorption centered around 1750 cm^{-1}, which is also shaded. The infrared absorption prevents a portion of the Earth's black body emission from being radiated into space. Without the atmospheric absorption, the Earth would be much colder.

The Reason Carbon Dioxide Is So Important

The reason carbon dioxide is so important can be seen by looking at the shaded regions of the black body spectrum and the absorption spectra. Water absorbs essentially everything at wavelengths longer than 500 cm^{-1}. However, the bottom two spectra in Figure 17.1 show that carbon dioxide absorbs strongly in the region where water absorption is not very strong. The carbon dioxide absorption is very close to the peak of the Earth's black body emission spectrum, as shown in the top part of Figure 17.1, regardless of which Earth surface temperature is used. Therefore, carbon dioxide strongly absorbs the Earth's black body radiation in an important spectral range where other components of the atmosphere, particularly water, do not. The carbon dioxide spectrum (middle panel of Figure 17.1) shows that there is close to zero transmission in the middle of the spectrum, approximately 667 cm^{-1}. However, as the concentration of CO_2 increases, the region of very strong absorption becomes wider and portions of the spectrum, where a few percent are transmitted, will transmit virtually nothing from the Earth's atmosphere into space. The net result is that CO_2 absorbs strongly near the peak of the Earth's black body spectrum where water doesn't, and an increase in CO_2 in the atmosphere will trap more black body radiation, thereby causing the planet to warm.

Why Carbon Dioxide Absorbs Where It Does

We see that carbon dioxide traps infrared light at the peak of the Earth's black body emission and that an increase in CO_2 concentration will have a deleterious effect on the Earth's temperature. But why does CO_2 absorb infrared light at wavelengths centered at 667 cm^{-1}? In Chapters 8 through 11, we discussed the energy levels of a particle in a box, of the hydrogen atom, and of all of the other atoms. In Chapters 12 to 14, we discussed molecular orbitals and the associated energy levels. All of these discussions concerned the energy levels associated with electrons. Using the ideas of molecular orbitals, the nature of bonds that hold atoms together to form molecules was explicated. What we have not discussed is the motions of atoms that are bonded together to form molecules.

Figure 12.1 displays the potential energy curve for the hydrogen molecule, H_2. The curve shows the energy at different separations of the hydrogen atom nuclei. The bond length is the separation where the energy is a minimum. However, the bond is not rigid. If we think about the bond using classical mechanics, the bond is a spring with two masses, the hydrogen atoms, attached at each end of the spring. A spring can be stretched and compressed. In a classical system, if you stretch the spring and let go, the masses will oscillate back and forth with the spring being alternately stretched and compressed. The masses of a classical oscillator will vibrate (oscillate) back and forth with a well-defined trajectory. Based on quantum theory, we should immediately suspect that a quantum vibration cannot have a well-defined trajectory. Such a trajectory would mean that we know the positions and moment of the particles (the atoms) precisely. Such knowledge for absolutely small systems, such as atoms bonded to form a molecule, violates the Heisenberg Uncertainty Principle.

Figure 17.2 shows a ball-and-stick model of carbon dioxide, CO_2, as well as representations of its possible vibrational motions. CO_2 is linear, with the two oxygens double bonded to the central carbon. CO_2 has four different vibrational motions, called vibrational modes. The bonds can stretch and compress as well as bend. The bonds are represented by springs. We will describe the motions as if they are classical balls connected by springs to understand the nature of the modes.

The Vibrational Modes of Carbon Dioxide

In the symmetric stretch, the central carbon does not move. As shown by the solid arrows, the two oxygens move away from the carbon, thereby stretching the springs. The two oxygens then move back toward the central carbon, compressing the springs, as indicated by the dashed arrows. For a classical ball-and-spring system, this motion is repeated, so the positions oscillate back and forth. The frequency of the oscillation is determined by the masses and the strengths of the springs. In the asymmetric stretching mode, the two oxygens move to the right. The oxygen on the right compresses the spring, and the oxygen on the left stretches the spring. A vibration does not move the molecule to a new location. Because both oxygens are moving to the right, the carbon moves to the left in order to keep the molecule in the same location. Because the carbon moves to the left when the oxygens move to the right, the average position of all of the mass, called the center of mass, is unchanged. The motions are indicated by the solid arrows. The direction of each atom then reverses, as shown by the dashed arrows.

The symmetric and asymmetric stretches maintain all three atoms on a line. In the bending mode, the two oxygens move up and the carbon moves down. This keeps the center of mass in one

FIGURE 17.2. *Top: Ball-and-stick model of carbon dioxide (CO₂). Bottom: The three different vibration motions that the molecule can undergo. There are two bending modes: the one shown and the equivalent one with the atoms going in and out of the plane of the page.*

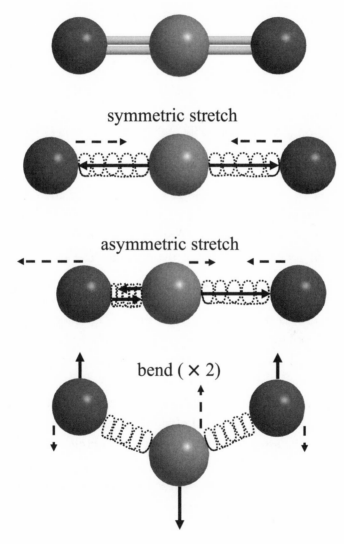

symmetric stretch

asymmetric stretch

bend (× 2)

place. Then the carbon moves up and the two oxygens move down. In addition to the bending mode shown in Figure 17.2, there is a second bending mode. The one shown has the motions of the atoms in the plane of the page. The second bending mode is identical except the atoms move into and out of the plane of the page.

Quantum Vibrations Have Discreet Energy Levels

In a classical vibrational oscillator made up of balls connected by springs, the energies the system can have are continuous. Consider the symmetric stretch. Three balls connected by two perfect springs are laying on a frictionless table with no air resistance. If you grab the outer two balls, stretch the springs the same amount, and let go, the balls will execute the symmetric stretching mode. Because the spring is perfect, the table is frictionless, and there is no air resistance (none of which is true in real life), the oscillation will continue forever. The period or frequency of the oscillation is independent of how far you stretch the springs. The period is determined by the springs' strengths and the masses. If you stretch the springs only a little bit, the balls will move slowly. Their average kinetic energy is small. If you stretch the springs a lot, the balls will move fast, and the average kinetic energy is large. The energy of the oscillating ball and spring system is continuous. It only depends on how much you stretch the springs.

Molecules are not really balls and springs. They are quantum mechanical systems composed of atoms joined by chemical bonds. Rather than having a continuous range of energies, the quantum system has discreet vibrational energy levels. The quantization of the energy is equivalent to the particle in a box problem discussed in Chapter 8. Gerhard Herzberg (1904–1999) won the Nobel Prize in Chemistry in 1971 "for his contributions to the knowledge of electronic structure and geometry of molecules, particularly free radicals." Herzberg's work on determining the structure of molecules was based to a large extent on his explanations for the vibrational spectra of molecules.

The energy of a classical racquetball is continuous, but the energy of the quantum racquetball (particle in a box) has energy levels (see Figure 8.6). Figure 17.3 shows a potential curve for a vibrational mode of a molecule, like the one shown in Figure 12.1, but now the first several quantized vibrational energy levels are also shown. Again like the particle in a box, the lowest energy level, n = 0, does not have zero energy.

Energies of Quantized Vibrations

The simplest model for the vibrational energy levels gives the energies as

$$E = h\nu(n + 1/2),$$

where h is Planck's constant, ν is the vibration frequency, and n is a quantum number that can take on values, 0, 1, 2, etc. For n = 0, the energy is $1/2 h\nu$. For n = 1, the energy is $3/2 h\nu$. So the difference in energy between the lowest energy level and the first excited vibration level is $h\nu$. In this model, all of the energy levels are equally spaced with a separation of $h\nu$. In real molecules, the energy levels get somewhat closer together as the quantum number increases. For our purposes, we only care about the spacing between the lowest energy level and first excited energy level.

CO_2 Bending Mode Absorbs at the Peak of the Earth's Black Body Spectrum

The bottom portion of Figure 17.3 shows the first two vibrational energy levels. Light will be absorbed at the energy of the separation of the levels, which is indicated by the dashed arrow. Since the difference in energy is $\Delta E = h\nu = c\, h/\lambda$, measurement of the light frequency (ν) or the wavelength (λ) at which light is absorbed gives

FIGURE 17.3. *Top: A potential energy curve showing the energy as a function of the bond lengths with the vibrational quantum levels. Only the first few energy levels are shown. Bottom: The lowest vibrational energy level (n = 0) and the first excited level (n = 1) for the CO_2 bending modes (Figure 17.2.). This transition (arrow) will absorb the Earth's black body radiation (see Figure 17.1).*

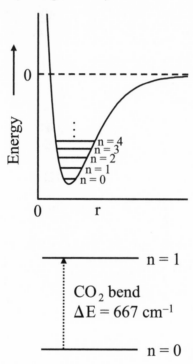

the oscillator frequency. As shown in the figure, $\Delta E = 667$ cm^{-1} for the bending modes of carbon dioxide. The two bending modes have the same frequency because they only differ by the direction of the bend. (We can write the energy or frequency in wave numbers [cm^{-1}] by dividing the energy, ΔE, by c h.) The frequency of light absorbed by the CO_2 bends is almost exactly at the peak of the Earth's black body spectrum. It is much easier (takes less energy) to bend a chemical bond than to stretch it. The carbon dioxide symmetric and asymmetric stretches are at much higher frequency. Neither contributes significantly to absorption of the Earth's black body radiation.

CO$_2$ GREENHOUSE EFFECT
IS QUANTUM MECHANICAL

The important point is that at the most fundamental level, CO$_2$'s contribution to the greenhouse effect and to global climate change is inherently quantum mechanical. First, the bonds that are broken and made in burning natural gas, oil, or coal are determined by the quantum mechanics that give us molecular orbitals, which control the bond strengths. The bond strengths determine the amount of energy that is released per CO$_2$ produced. At an even more fundamental level, the shape of the black body spectrum emitted by the Earth is determined by quantum effects. Black body radiation was discussed in Chapters 4 and 9. Planck's explanation of the shape of the black body spectrum and how it changes with the temperature of a hot object was the first application of quantum theory. The CO$_2$ absorption centered at 667 cm^{-1} is a result of the quantized vibrational energy levels of molecules, a purely quantum effect. The CO$_2$ bending modes have their quantized $n = 0$ to $n = 1$ vibrational transition at a key frequency in the Earth's black body spectrum. While massive power plants, vast numbers of cars, trucks, and planes, burning of rain forests, etc. produce the greenhouse gas CO$_2$, it is the quantum interaction between CO$_2$ and the Earth's infrared black body radiation that produces the greenhouse effect.

18

Aromatic Molecules

IN CHAPTERS 13 AND 14 we discussed double bonds, and in Chapter 16 we saw that double bonds play a fundamental role in determining the biological properties of fats. Some of the fat molecules discussed, the polyunsaturated fats, have several double bonds, but these double bonds are always separated by a number of single bonds. For example, Figure 16.5 shows a ball-and-stick model of DHA, a polyunsaturated fat with six double bonds. As can be seen, there are two single bonds between each double bond. In this chapter, we will see the wide-ranging impacts of multiple double bonds that are not separated by several single bonds. Quantum theory shows that the nature of the bonding found in the molecule benzene and a vast number of other "aromatic" molecules can explain electrical conductivity in metals, as well as the differences among metals, semiconductors, and insulators that will be discussed in Chapter 19. To understand aromatic molecules and electrical conductivity in metals, it is necessary to discuss the nature of the molecular orbitals that arise when identical atomic orbitals from many atoms interact to form MOs.

BENZENE: THE PROTOTYPICAL
AROMATIC MOLECULE

Figure 18.1 shows a diagram of benzene, which is composed of six carbon atoms and six hydrogen atoms. Experiments have determined that benzene is a perfect hexagon with all of the atoms, carbons, and hydrogens in a plane. The angle formed by the bonds from one carbon to its two nearest neighbors is exactly 120°, and the angle formed by the bond of a hydrogen to a carbon and that carbon's bond to an adjacent carbon is 120°. So the three bonds made by any carbon have a trigonal geometry, which means they are formed by the carbon using three sp^2 hybrid atomic orbitals on each carbon. That leaves one unused 2p orbital on each carbon, call them $2p_z$ orbitals, that point in and out of the plane of the page. We know that carbon always forms four bonds. Here, a carbon is only bonded to three other atoms, using up three bonds. The $2p_z$ orbital must be used to make a π double bond, but where is its location in the molecule?

FIGURE 18.1. *The geometry of benzene,C_6H_6. Benzene is planar, and it is a perfect hexagon.*

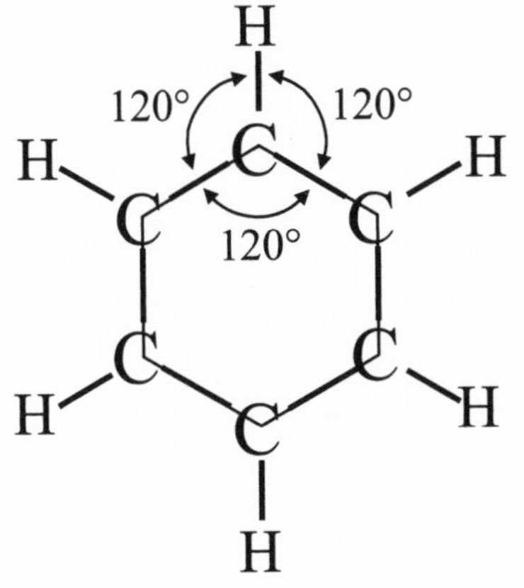

Where Are the Double Bonds?

Figure 18.2 shows two possible double-bonded structures. In both, each carbon makes four bonds. A carbon makes three σ bonds, one to a hydrogen and one to each of the adjacent two carbons. Each carbon also participates in a double bond with one adjacent carbon. The diagrams on the right and left are identical except for the locations of the double bonds.

Two things are wrong with the bonding in benzene depicted in Figure 18.2. In the discussion of double bonds in Chapter 14, Table 14.1 shows that the carbon-carbon double bond length is considerably shorter than the single bond length. In ethylene (double bond) vs. ethane (single bond), the bond lengths are 1.35 Å and 1.54 Å, respectively. So if benzene had alternating double and single bonds, it should have alternating short and long carbon-carbon bonds. However, experiments demonstrate conclusively that benzene is a perfect hexagon; all of the carbon-carbon bond lengths are the same.

If we ignore the diagram's implied unequal bond lengths, the second problem is the issue of which diagram would be correct, the

FIGURE 18.2. *Two possible configurations of double bonds in benzene. In both, each carbon makes four bonds.*

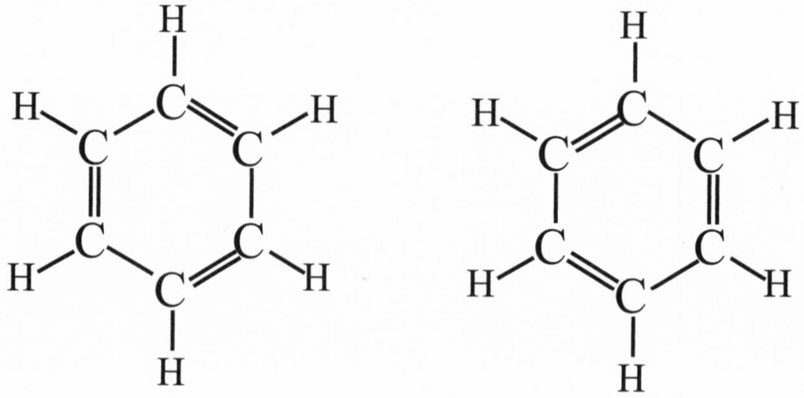

one on the right or the one on the left? There is no reason to prefer one over the other. An early explanation was that the bonds jumped back and forth between the configuration on the right and the configuration on the left. The jumps were proposed to produce a type of average structure. This idea is in some sense closer to the truth, but the real answer, which then applies to many types of systems, is in the nature of the molecular orbitals that are formed.

Delocalize Pi Bonds

Figure 18.3 is a schematic of the atomic orbitals used to form molecular orbitals in benzene. The top shows the atomic hybrid orbitals used to make the σ bonds. Each carbon uses three sp^2 hybrid orbitals to form three σ bonds, one to a hydrogen and one to each of the two adjacent carbon atoms. Forming the three sp^2 hybrid atomic orbitals leaves each carbon atom with a leftover p orbital. In the top part of Figure 18.3, take the plane that contains the atoms (the plane of the page) to be the xy plane. Then each of the carbons has an unused p_z orbital sticking in and out of the plane of the page. These orbitals are shown in the bottom of the figure. The positive and negative lobes of the p_z orbitals are above and below the plane of the ring. In the diagram, the lengths of the bonds between carbon atoms have been exaggerated and the widths of the p_z orbitals are reduced so that they can be seen clearly. The p_z orbitals actually overlap, as shown more to scale in Figure 14.15.

The six p_z atomic orbitals will combine to form molecular orbitals. The six atomic orbitals can contain at most 12 electrons without violating the Pauli Principle. Therefore, the six atomic orbitals will form superpositions to form six molecular orbitals that can also contain at most 12 electrons. The MOs are not associated with a particular atom or even a particular pair of atoms. They are molecular orbitals that span the entire six-carbon system.

In discussing the hydrogen molecule in connection with Figure

FIGURE 18.3. *Top: Benzene σ bonding. Each carbon makes three bonds using three sp² orbitals that lie in the xy plane. Each carbon has a p_z orbital perpendicular to the plane of the benzene ring. Bottom: The carbon p_z orbitals have positive and negative lobes that are above and below the plane of the ring. The bond lengths are exaggerated and the p_z lobes are made small for clarity of presentation. The lobes of adjacent p_z orbitals overlap.*

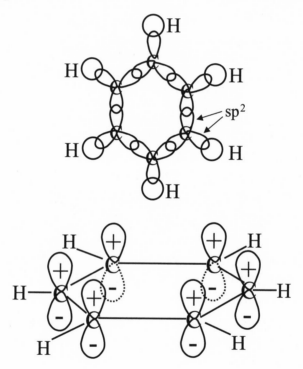

12.6, we saw that two atomic orbitals combine to form two molecular orbitals, one bonding MO and one antibonding MO. In Chapter 13, we investigated larger diatomic molecules, such as F_2, O_2, and N_2. In these atoms, three p orbitals on each atom, six atomic orbitals total, combined to form six π MOs, three bonding and three antibonding MOs (see Figure 13.5). Some of these diatomic π MOs were degenerate, that is, they have the same energy.

The Bonding and Antibonding Molecular Orbitals

In benzene, six p_z atomic orbitals combine to form three bonding MOs and three antibonding MOs, as shown in Figure 18.4. The six carbon $2p_z$ orbitals, one on each carbon atom, have identical energy. This is indicated by the six closely spaced lines on the left-hand side of Figure 18.4. These combine to form six MOs with energy levels shown on the right-hand side of the figure. Three of the MOs have energies lower than the p_z atomic orbital energy. These are the bonding MOs. Three of the MOs have energies higher than the atomic orbital energy. These are the antibonding MOs. Figure 18.5 shows the bonding and antibonding energy levels with the six electrons, one from each carbon, placed in the appropriate energy levels. We place the electrons in the lowest energy level consistent with the Pauli Principle. The Pauli Principle (Chapter 11) states that at most two electrons can be in a single orbital, and they must have opposite spin states (one up arrow and one down arrow). The first two electrons go into the lowest energy MO. The next two MOs

FIGURE 18.4. *Left: Benzene has six carbon atoms, each with a $2p_z$ orbital. These have identical energy, which is indicated by the six closely spaced lines. Right: The six p_z orbitals combine to form six π molecular orbitals, three bonding (b) and three antibonding (*) MOs.*

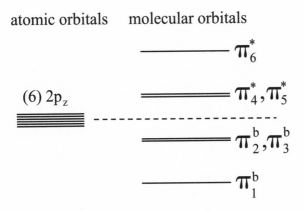

FIGURE 18.5. *Benzene π molecular orbital energy levels with the six electrons placed in the appropriate MOs in the lowest possible energy levels consistent with the Pauli Principle.*

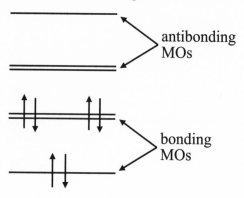

have the identical energy indicated by two closely spaced lines. Two electrons will go into each of these MOs. The three MOs filled by the six electrons are all π bonding MOs. The π antibonding MOs are empty.

The Carbon-Carbon Bond Order is 1.5

Figure 18.5 shows that the six carbon p_z electrons occupy three π bonding MOs. Therefore, there are three π bonds shared by six carbons. These three π bonds are in addition to the σ bonds that connect each carbon to its two nearest carbon neighbors. The net result is that each carbon has 1.5 bonds to other carbons. The three π bonds shared by the six carbon atoms contribute a half of a bond joining adjacent carbons. The bonds between carbon atoms are shorter and stronger than a carbon-carbon single bond, but not as short or as strong as a full double bond. The π bonding keeps the molecule rigorously planar. Twisting the ring away from planarity reduces the overlap of the p_z orbitals and raises the energy. Figure 18.6 shows a chemical diagram of benzene. A carbon is represented

FIGURE 18.6. *Benzene chemical diagram. A carbon atom is at each vertex, and a hydrogen is at the end of each line from a carbon. The circle represents the delocalized π molecular orbitals.*

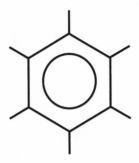

by a vertex. A hydrogen atom is at the end of each line that emanates from a carbon. The circle represents the delocalized π electron system.

Many molecules have carbon rings with delocalized π bonding. Another example is naphthalene, which is shown in Figure 18.7. Naphthalene has 10 carbons forming two six-membered rings with eight hydrogens. The two circles represent the delocalized π molec-

FIGURE 18.7. *Naphthalene chemical diagram. Naphthalene has 10 carbons and eight hydrogens. The circles represent the delocalized π molecular orbitals.*

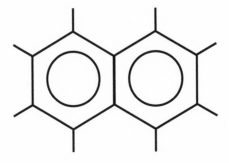

ular orbitals. Like benzene, naphthalene is planar and each carbon has 1.5 bonds to the adjacent carbons.

THE BENZENE DELOCALIZED
PI MOLECULAR ORBITALS

Benzene, naphthalene, and similar molecules are referred to as aromatic molecules. They tend to have a sweet smell. Naphthalene is a mothball, which has a characteristic aromatic smell. Perfumes are more complex aromatic molecules having a number of benzene-like rings, as well as other chemical groups replacing the hydrogen atoms. Small changes in the molecular structure change the aroma, which is what makes one perfume smell different from another.

In Chapter 8, we discussed the particle in a box problem. Figures 8.4 to 8.6 show the particle in a box wavefunctions and energy levels. The wavefunction associated with the lowest energy level has no nodes. The next higher energy state has a wavefunction with one node, the next higher energy state has a wavefunction with two nodes, and so forth. A node is a place where the wavefunction goes to zero, so the probability of finding the electron is zero. The particle in a box is a one-dimensional problem. A node is a point. In Chapter 10, we examined the wavefunctions and the energy levels for a hydrogen atom. Figures 10.2 to 10.6 show representations of the wavefunctions for the 1s, 2s, and 3s hydrogen atom states. The hydrogen atom wavefunctions are three-dimensional. The lowest energy state (1s) has no nodes, the next higher energy state (2s) has a wavefunction with one node. The next higher energy state (3s) has two nodes. These nodes are three dimensional surfaces on which there is zero probability of finding the electron.

The benzene π MOs also have an increasing number of nodes as the energy is increased. Figure 18.8 shows schematics of the benzene π MOs. The shaded areas are the regions of high electron density (high probability of finding electrons) for the π MOs.

FIGURE 18.8. *Benzene π molecular orbital energy levels and schematics showing the shapes of the corresponding MOs. As the energy increases, the number of nodes increases. MOs with the same number of nodes have the same energy.*

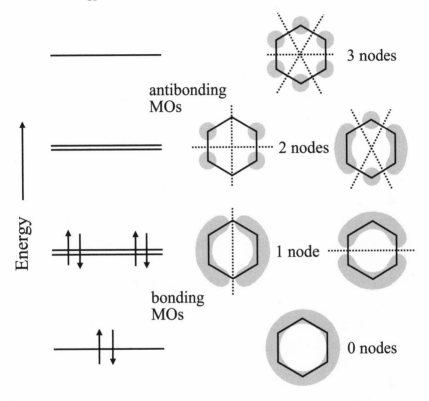

These are three-dimensional electron clouds that extend above and below the plane of the page and do not have sharp boundaries. Also shown are the energy levels with the 6 p_z electrons filled in the lowest energy bonding MOs. The lowest energy MO has no nodes. There are two states with the next higher energy. Both of these MOs have one node. The three MOs with no nodes and one node are the bonding MOs. There are also two levels with the next higher energy. These MOs have two nodes. The highest energy MO has three nodes. The three MOs having two and three nodes are the antibonding MOs.

Comparing the schematics of the lowest energy MO with the highest energy MO in Figure 18.8, it can be seen clearly why the former is a bonding MO and the latter is an antibonding MO. The lowest bonding MO has electron density between all of the carbons. The highest antibonding MO has nodes between all of the carbons, so electrons in this MO would not join the carbons together. The pair of bonding MOs, although higher in energy than the lowest energy bonding MO, results in bonding among the carbon atoms. Each of these MOs has one node. The one on the left puts electron density between a pair of carbons on the left and right. The MO on the right puts electron density between three carbon atoms on the top and three on the bottom. In spite of the nodes, these MOs combine with the lowest energy MO to produce the three π bonds that are shared by the six carbons. The two degenerate antibonding MOs have two nodes each. The one on the left clearly does not contribute to bonding because it does not place electron density between any of the carbons. The one on the right puts electron density between two pairs of carbons, but when combined with the one on the right, it does not produce net bonding.

Light Absorption by Aromatic Molecules

Quantum theory can calculate the molecular orbitals of aromatic molecules, as well as their shapes and sizes. There are many ways to test the quantum calculations by comparison to experiments. One of the most useful is to employ optical spectroscopy to measure the wavelengths (colors) of light absorbed by a molecule. We will use naphthalene as an example.

Figure 18.7 shows a diagram of naphthalene, with its 10 carbon atoms. Each carbon atom will contribute one p_z orbital with one electron to form the delocalized π electron system. The other three carbon valence electrons are used to form the σ bonds. Ten p_z atomic orbitals form the π system, so there will be 10 molecular orbitals, five bonding MOs and five antibonding MOs. In naphthalene, none of the MOs is degenerate; each has a different energy. Figure 18.9 displays a schematic of the energy levels of the π MOs of naphthalene. The left-hand side shows the π energy levels with the 10 π electrons filling the five bonding MOs. The antibonding MOs are empty.

The right side of Figure 18.9 illustrates the effect of absorption of light. Because the energy levels are quantized, only certain energies of light can be absorbed by the molecule. In the figure, ΔE is the lowest energy light that can be absorbed. Consider what happens if we shine light on a sample of naphthalene molecules, beginning with light that is too low in energy to be absorbed by the

FIGURE 18.9. *A schematic of the naphthalene π molecular orbital energy levels. There are five bonding MOs and five antibonding MOs. The left side shows the 10 p electrons filling the five bonding MOs. The right side shows the result of absorption of light that raises the energy of a bonding electron into an antibonding MO.*

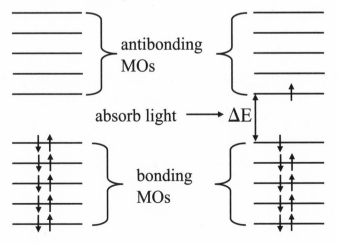

antibonding MOs

absorb light ⟶ ΔE

bonding MOs

molecules. The energy of the light is $E = h\nu$, where h is Planck's constant and ν is the frequency. So initially, $\Delta E > h\nu$; the energy separation between the highest occupied molecular orbital (HOMO) and the lowest unoccupied molecular orbital (LUMO) is greater than the energy of the photons that are impinging on the sample. As a result, they pass through the sample without being absorbed. Now, we start changing the energy of the light to higher energy (red to blue). When $h\nu = \Delta E$, light is absorbed, which is detected by a decrease in the amount of light that passes through the sample. An electron is excited from the HOMO to the LUMO. This excitation is shown on the right side of Figure 18.9, which has one electron in the HOMO and one electron in the LUMO. In the left-hand side of Figure 18.9, there are two electrons in the HOMO and none in the LUMO.

The transition from the HOMO to the LUMO is the lowest energy transition. As drawn, the bonding MOs are closer together than the separation between the HOMO and the LUMO. However, an electron cannot be excited from one filled bonding MO to another filled bonding MO. If we try to take an electron from one bonding MO and put it into another bonding MO, the result would be an MO with three electrons in it. Three electrons in an MO violates the Pauli Exclusion Principle. Then in our optical absorption experiment, as the light color is scanned from red to blue (low energy to high energy), the first color (wavelength) that is absorbed corresponds to the energy ΔE. ΔE can be calculated using quantum mechanics. ΔE depends on the structure of the molecule and the interactions of the atomic orbitals that give rise to the molecular orbitals. The quantum calculations can be tested further by comparison to the wavelengths at which absorption of higher energy light takes place as the light is scanned further and further to the blue. This second absorption occurs because light can take an electron from the HOMO to one energy level above the LUMO. Another absorption at even higher energy will occur by exciting an electron from the HOMO to two levels above the LUMO, and so forth.

NAPHTHALENE TREATED AS A PARTICLE IN A BOX PROBLEM

Using modern quantum theory and computers, the structure of naphthalene can be calculated with great accuracy. The theory will give the bond lengths and bond angles. For example, the bond lengths can be calculated to 0.001 nm, that is, a thousandth of a nanometer. The calculations also yield the frequencies at which light is absorbed with substantial accuracy. The calculations use the masses, number of electrons, and charges of the nuclei. The calculations will include both the σ and π bonding. As we have discussed, the π electrons are not localized on one or two carbon centers, but rather are delocalized over the entire carbon framework of the molecule. The lowest energy absorption of naphthalene, the HOMO to LUMO absorption, occurs at 320 nm, which is in the ultraviolet portion of the optical spectrum.

We can do a very crude calculation by pretending that the π electrons are particles in a box. In Chapter 8, we discussed the particle in a box in great detail. If we take the HOMO to LUMO transition to be a transition of an electron in a box from the $n = 1$ to $n = 2$ levels (see Figure 8.7), we can use the formulas derived just below Figure 8.7. For this transition we found that

$$\Delta E = \frac{3h^2}{8mL^2}$$

where h is Planck's constant, m is the mass of the electron, and L is the length of the box. Here we will take L to be 0.51 nm, the distance across the carbon framework of naphthalene. Then,

$$\Delta E = \frac{3(6.6 \times 10^{-34})^2}{8(9.1 \times 10^{-31})(0.8 \times 10^{-9})^2} = 6.9 \times 10^{-19} \text{ J}.$$

Converting this energy to a frequency by dividing by h gives $v = 1.04 \times 10^{15}$ Hz. Then the wavelength of the light that will be absorbed is $\lambda = 2.87 \times 10^{-7}$ m $= 287$ nm. This wavelength is further in the ultraviolet than the true absorption, but not that far from the observed value.

The particle in a box calculation shows that if a particle with the mass of an electron is confined to a box the size of naphthalene, the first absorption will be in the ultraviolet range. The reasonable accuracy of the particle in a box calculation for naphthalene is somewhat fortuitous. Even if we wanted to model naphthalene as a particle in a box, it should be a two- or three-dimensional box, not a one-dimensional box. Such calculations produce results with significant errors. However, an accurate quantum theory calculation will yield the molecular structure and much more accurate frequencies for absorption of light. In addition, if, for example, a hydrogen is replaced by a fluorine, quantum theory will accurately predict how much the frequency of light absorption of fluoronaphthalene is changed from that of naphthalene.

19

Metals, Insulators, and Semiconductors

FIGURE 19.1 IS A SCHEMATIC DIAGRAM of a battery connected to a metal rod. We will discuss sodium metal below as an example, but the rod could be any metal. The positive end of the battery pulls electrons from the metal rod. Electrons flow out of the rod into the battery.

FIGURE 19.1. *A metal rod of sodium, for example, is connected by wires to a battery. Negatively charged electrons are pulled from the metal rod to the positive side of the battery. Electrons flow from the negative side of the battery back into the metal rod.*

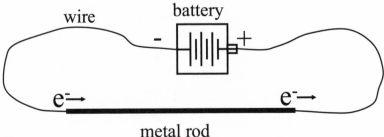

metal rod

However, to keep the rod from developing a positive charge that would pull back on the electrons and stop the flow, the rod must be connected to the negative side of the battery.

Electrons flow from the negative side of the battery into the rod, keeping the rod neutral, that is, the rod does not develop an electrical charge. Instead of a rod, the electrons could flow through the filament of a lightbulb in a flashlight. The flow of electrons through the filament causes it to get very hot and produce black body radiation in the visible part of the spectrum.

METALS

Delocalize Molecular Orbital for a Metal

How can electrons move through a piece of metal? What is the difference between a metal and an insulator? What is a semiconductor? Why does a metal get hot when electrons flow through it? What is superconductivity? To answer the first three questions, we will extend our discussion of the types of delocalized molecular orbitals found in aromatic molecules such as benzene and naphthalene (see Chapter 18) to the MOs of a macroscopic piece of metal and other materials. To answer the last two questions, we will expand the discussion to the effect of the thermal vibrations of the atoms that make up a piece of metal on electron motion in a metal.

In Chapter 10 on the hydrogen molecule, we saw two hydrogen atomic orbitals combine to form two molecular orbitals, one bonding and one antibonding. In benzene, we saw that six p_z atomic orbitals, one from each carbon, combined to form six MOs, three bonding MOs and three antibonding MOs. For naphthalene, 10 p_z atomic orbitals combined to form 10 MOs, five bonding and five antibonding. In each case, the MOs span the entire molecule. In Chapter 11 on the Periodic Table of Elements, we said that sodium, Na, is a metal because it has one electron, the 3s, past the neon filled shell configuration. Na can readily give up an electron to form salts, such as table salt, NaCl. In water NaCl dissolves to become Na^+ and Cl^-. We said that Na as a solid was a metal and conducted electricity. Now we are in a position to see why.

First consider the 3s orbitals of two sodium atoms that are next to each other and interacting. For sodium, the 3s electron is the valence electron, which will participate in bonding. The top portion of Figure 19.2 shows the energy levels of the two 3s atomic orbitals combining to form MOs. One of the MOs has lower energy than the atomic orbitals. It is the bonding MO. The other MO has higher energy; it is the antibonding MO. The middle shows that three atomic orbitals will form three MOs. The bottom illustrates the situation for six interacting sodium atoms. The six 3s atomic orbitals combine to form six MOs, three bonding and three antibonding.

Each Na has one 3s electron, which it will contribute to fill the MOs. For the system with six sodium atoms, there will be six elec-

FIGURE 19.2. *Top: Two sodium 3s atomic orbitals interact to produce two molecular orbitals, one lower in energy (bonding) and one higher in energy (antibonding). Middle: Three 3s atomic orbitals interact to form three MOs. Bottom: Six 3s atomic orbitals interact to form six MOs.*

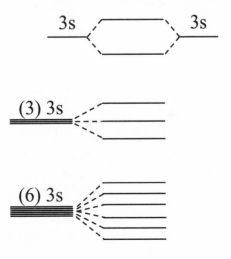

trons to fill the MOs. Each MO can be occupied by two electrons with opposite spins (one up arrow and one down arrow). Therefore, the three lowest energy MOs, which are the bonding MOs, will be filled by the six electrons. The three higher energy MOs are empty.

Now we must consider what happens when we have a very large number of interacting sodium atoms. Consider a metal rod composed of sodium atoms that is 10 cm long and 1 mm in diameter, such as that in Figure 19.1. For the dimensions given, the rod contains $N = 2 \times 10^{21}$ Na atoms, where N is the number of atoms. This number is two billion trillion atoms. The two billion trillion 3s atomic orbitals combine to form two billion trillion molecular orbitals. Like the MOs of benzene or naphthalene, the MOs of the sodium rod should be thought of as spanning the entire system, that is, the entire piece of metal.

A Piece of Metal Has a Vast Number of MO Energy Levels Called a Band

Figure 19.3 illustrates the energy levels of this system. Each of the N sodium atoms has an electron in a 3s atomic orbital. In the absence of interactions between the atoms, all of these atomic orbitals have the same energy. In the figure, this is represented by the collection of closely spaced lines on the left-hand side. To show that there are many atomic levels, the lines have been spread out, but they all have the same energy. When the atoms interact, the N atomic orbitals form N MOs. As we have seen previously for molecules, the MOs have different energies. Some of the MOs have lower energies than the atomic orbital energy and some have higher energy. This is represented on the right side of the figure by the spread-out but closely spaced lines. The MO energy levels in Figure 19.3 are equivalent to those shown in Figures 18.8, 18.9, and 19.2, except that there are vastly more energy levels, which are much more closely spaced. These are referred to as a band of states.

FIGURE 19.3. *In a piece of sodium metal, there are N atoms. Each has an electron in a 3s orbital, represented by the closely spaced lines on the left. These all have the same energy. The N 3s atomic orbitals interact to form N molecular orbitals with energy levels shown on the right. The MO energy levels are so close together that the energy is effectively a continuous band of states. The Fermi level marks the highest occupied molecular orbital.*

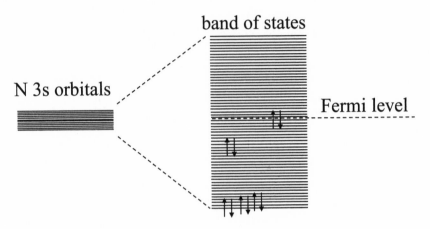

Quantum theory shows that the width of the band of states, that is, the difference in energy between the highest energy MO and the lowest energy MO, is only a few times the energy splitting of the MOs that arise from a pair of interacting sodium atoms (see Figure 19.2, top). Then in our example of two billion trillion Na atoms, there are this many energy levels in a relatively narrow range of energy. The result is that the energy levels are so closely spaced that the energy is effectively continuous within the band.

Putting in the Electrons

There are N sodium atoms, each with a single 3s electron. We need to take these N electrons and put them in the appropriate MOs, as we did with small molecules in Chapters 12 and 13 and as shown in Figures 18.8 and 18.9. These sodium metal delocalized MOs are orbitals like any others, so we must obey the three rules for putting in the electrons discussed in Chapter 11. They are lowest energy first, no more than two electrons in an orbital that must have paired spins (Pauli Exclusion Principle), and don't pair spins unless necessary (Hund's Rule). Figure 19.3 illustrates putting in the electrons. The first electron goes in the lowest energy level. The next electron goes into the same level with the opposite spin, that is, one up arrow and one down arrow. The third electron can't go into the lowest energy level because that would violate the Pauli Principle. So, it goes into the level one up from the lowest. The fourth electron goes into this same level with paired spins. This will continue until all N electrons are in MOs.

The Fermi Level

There are N MO energy levels and N electrons. But two electrons can go in each level. Therefore, only the bottom half of the energy band of levels will be filled. This is like benzene (Figure 18.8) and naphthalene (Figure 18.9), which also only have the bottom half of their MOs filled. The energy of the highest filled orbital is called the Fermi level, for Enrico Fermi (1901–1954). Fermi was a physicist who worked in many areas of science including the theory of solids, such as metals, and the theory of nuclear reactions. His work contributed to the development of nuclear energy. He won the Nobel Prize in Physics in 1938 "for his demonstrations of the existence of new radioactive elements produced by neutron irradiation, and for his related discovery of nuclear reactions brought about by slow neutrons." As we will see, the Fermi level is very important.

The Fermi level is the level of the highest filled MO at the absolute zero of temperature. This temperature is $0°$ K, where K is the Kelvin unit of temperature. One degree K is the same as one degree C (centigrade) except the scale begins at the absolute zero of temperature. $0°$ K is $-273°$ C or $-459°$ F. We have briefly discussed how heat in systems of molecules, such as water, causes the molecules to jiggle around. In Chapter 15, it was pointed out that the thermal motions of water were responsible for the breaking of hydrogen bonds between water molecules. As temperature is decreased, there is less and less heat (thermal energy) and the motions of atoms and molecules decrease. The absolute zero of temperature, $0°$ K, is the temperature at which there is no heat to cause atoms and molecules to move. The Fermi level is actually defined to be the energy of the highest filled MO at $0°$ K.

How Electrons Move Through Metal

As shown in Figure 19.1, electrons enter one side of the metal rod and leave the other. This is possible because the electrons are in delocalized MOs that span the entire piece of metal. However, quantum theory shows that if all of the electrons occupy only the MOs below the Fermi level, the electrons will not move in a particular direction. Real metals are three dimensional, but for this discussion let's consider only one dimension at a time. In our metal rod, when it is not connected to the battery, the electrons in the MOs are, nonetheless, constantly moving. Although the electrons are described in terms of quantum mechanical wavefunctions, the electrons have kinetic energy. Therefore, it is possible to calculate an electron velocity. The electrons in some MOs can be thought of as moving to the right. There are corresponding MOs of identical energy with electrons moving to the left. With all of the MOs filled, as shown in Figure 19.3, there will be no net electron current because there are an equal number of electrons moving to the right and to the left. In three dimensions, for any direction you pick, there will be equal probability of electrons moving in that direction and in the exact opposite direction.

However, when the metal rod in Figure 19.1 is connected to a battery, things change. One end of the rod is connected to the positive side of the battery and the other end is connected to the negative end of the battery. The connection to the battery changes what the electrons feel. Without the battery, the electrons feel the positive charges of the sodium nuclei and the negative charges of other electrons. Any one electron in the middle of the rod sees no difference between right and left. But with the battery there is an additional influence, the electric field produced across the metal rod by the battery. The electrons are attracted to the positive end and repelled from the negative end. The effect is to modify the system with the result that some electrons are in levels above the Fermi level that existed without the battery (see Figure 19.4). The electron states of the system are changed such that there are more electrons moving in the direction toward the positive end of the metal rod than toward the negative end.

FIGURE 19.4. *Schematic of the sodium metal 3s band of levels as shown in Figure 19.3, but now with the influence of being connected to the battery. The effect is to put some electrons above the no battery Fermi level, taking them from filled MOs to empty MOs. These electrons are represented by the arrows above the Fermi level.*

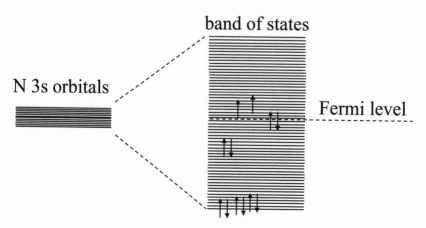

Quantum theory shows that it is necessary to have electrons above the Fermi level for electron conduction to occur. Because there are only infinitesimal differences in energy between the levels, even a very low voltage applied to the rod, which produces a tiny electric field, is sufficient to put some electrons above the Fermi level. The result is an electrical current flowing throw the metal rod. Electrons leave the positive end of the rod and are replaced by electrons entering from the negative end. For a bigger electric field (higher voltage), more electrons will be above the zero field Fermi level, and the electrical current is bigger. The detailed quantum theory of electrical conductivity in metals says that current will flow when an electric field is applied even at absolute zero temperature. Heat is not required for a metal to conduct. We will see below that this is not the case for semiconductors, and also, that heat, which is present at all temperatures above $0°$ K, actually interferes with electrical conductivity in metals.

INSULATORS

Insulators Do Not Conduct Electricity Because the Band Is Filled

Metals conduct electricity easily even at $0°$ K because the electrons only fill part of the band of states, as shown in Figure 19.3 and 19.4. A very small electric field (voltage) will put electrons above the Fermi level. An insulator is a material, such as glass or plastic, that does not conduct electricity at any temperature. A schematic illustration of the band structure of an insulator is shown in Figure 19.5. In sodium metal, the 3s electrons are the valence electrons. The valence band is only half filled. In an insulator, like quartz (SiO_2, silicon dioxide), which is very similar to glass, sharing of the electrons completes the shell of electrons. The interactions in a quartz crystal produce a band of states, with delocalized MOs, like in a metal. However, the valence band is completely filled. There are two

FIGURE **19.5.** *Schematic of the band structure of an insulator. There is a filled band, with two electrons in each MO. At much higher energy, there is an empty band.*

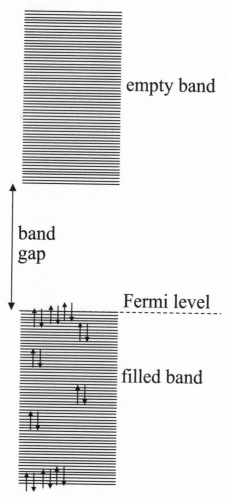

electrons in each MO because there are N MOs but 2N electrons. So, all of the MOs have two electrons, from the lowest energy level to the highest energy level in the band. The filled band is shown in Figure 19.5 by the presence of arrows from the bottom of the band to the top of the band. This filled band should be compared to the half-filled band of Na metal shown in Figure 19.3.

The Band Gap Is Large in Insulators

There are empty atomic orbitals at much higher energy, and these form MOs. However, there are no electrons to put in these MOs. Therefore, the next higher energy band is completely empty. The difference in energy between the top of the filled band and the bottom of the unfilled band is called the band gap. The Fermi level is at the top of the filled band.

As discussed qualitatively above and shown in detail by quantum theory, conductivity requires electrons in MOs with energies above the Fermi level. When an electric field is applied across the material (connected to a battery or other source of voltage), the nature of the delocalized states shifts. In a metal, because the band is only half filled and the energy levels are only separated by an infinitesimal amount, an applied electric field will result in a change such that some electrons are above the zero field Fermi level, and electrons flow through the metal. In an insulator, the next level above the Fermi level is in the empty band. The band gap is large, and the application of an electric field cannot change the system enough to put electrons into the empty band. Therefore, application of an electric field to an insulator is insufficient to produce conductivity, in contrast to a metal.

Another possibility is that thermal energy could excite electrons from the filled band to the empty band. An insulator has the property that the band gap energy is much greater than the thermal energy. As the temperature is increased, the amount of thermal energy increases. But an insulator has a band gap that is so large that the insulating material will be destroyed at temperatures that are still insufficient to thermally excite electrons from the filled band to the empty band. The net result is that application of an electric field cannot modify the states in a way to produce conductivity and thermal excitation of electrons cannot occur. Therefore, insulators do not conduct electricity.

SEMICONDUCTORS

In a Semiconductor the Band Gap Is Small

A semiconductor is like an insulator, but with a small band gap. A schematic of the band structure of a semiconductor is shown in Figure 19.6. In a semiconductor, such as silicon (Si), there are enough electrons to fill the valence band completely. At 0° K, where there is no thermal energy to excite electrons, all of the electrons are paired in the valence band. The Fermi level is at the top of the filled valence band. Therefore, silicon is an insulator at 0° K. However, in silicon and other semiconductors, the band gap is small. At room temperature, there is sufficient thermal energy to excite some electrons above the Fermi level into the next band. The thermal energy is contained in the motions of the atoms in a piece of semiconductor. The thermal excitation of electrons above the Fermi level into the next band is illustrated in Figure 19.6. The electrons that have been excited from filled MOs of the valence band to empty MOs of the conduction band are represented in the figure by arrows above the Fermi level. Because there are electrons above the Fermi level, a piece of semiconductor like silicon can conduct electricity. The electrons in the conduction band are called the conduction electrons.

Semiconductors do not conduct electricity as well as metals because they have far fewer conduction electrons. In a metal, there is no band gap. Large numbers of electrons are easily promoted above the Fermi level. In a semiconductor, there is a band gap, but it is small enough that thermal energy can excite some electrons above the Fermi level into the conduction band. As the temperature of a semiconductor is reduced, there are fewer and fewer conduction

FIGURE 19.6. *Schematic of the band structure of a semiconductor. The valence band is essentially filled. The gap in energy to the next band is relatively small. Some electrons are thermally excited above the Fermi level into the conduction band.*

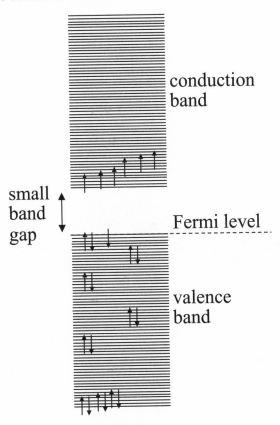

electrons to carry electrical current. At sufficiently low temperature, semiconductors become insulators. The only difference between a semiconductor and an insulator is the size of the band gap. Your computer chips, which are composed mainly of silicon semiconductors, will not work if they are sufficiently cold. The computers and electronics on a satellite must be kept warm or they will cease to function.

Thermal Energy Affects Electrical Conduction in Metals

Thermal energy is necessary in semiconductors to generate conduction electrons. Thermal energy also strongly influences electrical conduction in metals, although thermal energy is not necessary to generate the conduction electrons. In a piece of metal wire connected to a battery, there are electrons moving toward the positive end. As electrons leave the wire, they are replenished by electrons entering from the negative side of the battery. Current (electrons) flowing through a piece of wire will cause its temperature to rise. The heating elements in an electric stove or an electric space heater get very hot from a large current flowing through them. They get so hot that they glow red. The red color is black body radiation from the hot metal. We have said that electrons can readily flow through a piece of metal because the electrons are in delocalized MOs that span the metal. It only requires an electric field (connection to a battery or other voltage source) to get them moving in a particular direction. So the question is why does the flow of electrons cause the metal to heat up?

The electrons in a metal should be thought of as wave packets that are more or less localized. We discussed wave packets in Chapter 6 in connection with the Heisenberg Uncertainty Principle. The electron wave packets in a metal are formed from superpositions of the delocalized electronic MO wavefunctions in a manner analogous to photon wave packets or electron wave packets in a vacuum that are superpositions of the delocalized momentum states. Electrons are negatively charged so an electron wave packet carries a negative charge. The electron is accelerated toward the positive end. The acceleration gives the electron increased kinetic energy.

Vibrations of a Solid Are Phonons

In Chapter 17, we briefly discussed the quantized vibrations of molecules in connection with the greenhouse gas, carbon dioxide. A

piece of metal, which is made up of atoms, also has quantized vibrations. The atoms in a metal crystal lattice can jiggle around in their
positions. Although they jiggle, an atom remains in the same spot
on average. The motion of each atom is connected to the motions
of the other atoms in the same manner that the motions of the
atoms of a CO_2 molecule are connected to each other (see Figure
17.2). CO_2 has several distinct vibrations, symmetric and antisym-
metric stretches, and two bending modes. These three different
types of modes have vibrational energies (frequencies) that are very
different from each other.

In a metal crystal lattice, each atom can move in all three dimensions. For N atoms, there are 3N lattice vibrations, where again, N
is the number of atoms in the piece of metal. For any finite size
piece of metal, this huge number of vibrations results in a band of
vibrations rather than several discreet frequencies. At low temperature, only the lowest energy part of the band of vibrations is thermally excited. At higher temperature more lattice vibrations are
excited and higher energy vibrations are excited. The excited vibrations have kinetic energy. The energy of the excited vibrations is
what we think of as heat.

The quantized vibrations of a lattice are called phonons. This
name came about because phonons in certain fundamental quantum theory properties bear some resemblance to photons. Each
phonon is a delocalized vibrational wave that spans the entire crystal
lattice. The lattice waves can form more or less localized wave packets through the superposition of a range of wavelengths of lattice
waves. The more or less localized phonon wave packet is completely
analogous to the photon and electron wave packets mentioned just
above and discussed in Chapter 6. The phonons are moving wave
packets of mechanical thermal energy. The phonon wave packet can
be thought of as a moving region of relatively localized jiggling of
the atoms.

Electron Wave Packets and Phonon Wave Packets Scatter

An electron wave packet that is being accelerated toward the positive direction can interact with a phonon. The phonon causes the positively charged atomic nuclei to move. The negatively charged electron is influenced by the moving positive charges. The interaction of the electron and phonon is called a scattering event and is shown schematically in Figure 19.7. The electron and phonon wave packets are propagating in certain directions. The electron is being accelerated by the electric field when it "collides" with a phonon. Following

FIGURE 19.7. *Cartoon of an electron-phonon scattering event. The interaction of the electron and phonon causes the directions of the wave packet motions to change.*

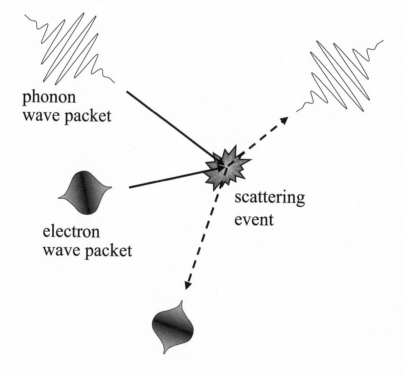

phonon
wave packet

scattering
event

electron
wave packet

the scattering event, in general, both wave packets will move in new directions. The electron will again be accelerated by the electric field toward positive. After some time, it will again encounter a phonon, and scatter. Each time the electron scatters, it gives up some of its kinetic energy to a phonon that it got from being accelerated by the electric field (voltage source).

The scattering events do two things. First, they prevent the electrons from moving directly to the positive battery connection. Second, the scattering events add kinetic energy to the phonons. The electron loses energy, and the phonon gains it. The electron-phonon scattering reduces the electrical conductivity of metals because the electrons keep getting bumped, which causes them to move in different directions as they are trying to move to the electrically positive end of the wire. This is called electrical resistance. At very low temperature, there are few phonons, so the electrons can move a long way between scattering events. This makes it easier for the electrons to reach the positive connection. As the temperature is increased, there are more and more phonons because phonons are heat. As the temperature goes up, the electrons propagate less distance before their direction is changed, reducing their ability to reach the positive electrode. The result is that electrical conductivity decreases (the resistance increases) as the temperature increases.

Electron-Phonon Scattering Heats the Metal

Because the scattering events add kinetic energy to the phonons, they raise the metal's temperature. Temperature is a measure of the heat content of a piece of material. Heat is the kinetic energy of the atomic motions. If there are a lot of electrons moving through the metal undergoing scattering events, then a lot of heat is added to the wire, and the temperature goes up. However, when the temperature goes up, there are even more phonons, more scattering events, and so the temperature goes up more.

This process is what you see when you turn an electric stove to high, and it takes some time for the heating element to glow red. When you first turn on the stove, the heating wire is at room temperature. Current (electrons) is flowing and electron-phonon scattering is occurring, causing the temperature to go up. The increased temperature means there are more phonons and more scattering events, and even more heat is added to the wire. The temperature goes up even more. However, as the temperature goes up, the current goes down because the additional scattering slows the progress of the electrons through the wire. The wire will come to a constant high temperature determined by the amount of current (the setting of the knob on the stove) that was initially applied at room temperature when the stove was first turned on.

Electrons in a normal metal will undergo electron-phonon scattering at any finite temperature. Therefore, a piece of wire has electrical resistance at any temperature other than absolute zero, $0°$ K. At absolute zero, there is no heat, so there are no phonons. However, it is impossible to reach absolute zero. To cool something down, you need something colder to take heat away. It is possible to achieve very low temperatures, for example, one millionth of a degree above absolute zero, using very specialized experimental methods, but even at this unbelievably low temperature, there are still some phonons and some electron-phonon scattering. In addition, if a piece of normal wire is at very low temperature, and you flow a significant amount of current through it, it will heat up. As mentioned in Chapter 17, electrical transmission lines from power plants to cities lose a lot of electricity. We now see why that is. It is caused by the electrical resistance of the wire, that is, electron-phonon scattering.

SUPERCONDUCTIVITY

Materials that have no electrical resistance at finite temperature are called superconductors, and the flow of electrons through a superconducting piece of wire is called superconductivity. In metals, superconductivity only occurs at very low temperature. The Dutch physicist Heike Kamerlingh Onnes (1853–1926) discovered superconductivity in 1911, when he cooled mercury metal to 4° K (− 269° C, − 452° F). He observed that the resistance went to zero. Some other metals and the highest temperature at which they are superconducting are niobium: 9.26° K, lead: 7.19° K, vanadium: 5.3° K, aluminum: 1.2° K, and zinc: 0.88° K. Superconductivity was not explained until decades later. In 1972, three American physicists, John Bardeen (1908–1991), Leon Cooper (1930–), and John Schrieffer (1931–) won the Nobel Prize in Physics "for their jointly developed theory of superconductivity, usually called the BCS–theory." The BCS Theory, developed in 1957, is a detailed quantum mechanics description of electron-phonon interactions at low temperature. In 1956, Leon Cooper showed that electron-phonon interactions can cause electrons to pair. Two electrons will in some sense be joined together even though they are physically far apart. BCS used this idea to show that these Cooper pairs do not undergo electron-phonon scattering of the type discussed above that leads to electrical resistance. When electron-phonon scattering is absent, the electrons move through the metal with no resistance even though the temperature is not absolute zero. Since there is no resistance, there is no loss of electrical energy in spite of the fact that a large current may be flowing.

Superconductors are used today in a variety of applications, and there is promise for very important widespread applications in the future. Magnetic resonance imaging (MRI) requires a very large magnet. The large cylinder that an MRI subject is placed in is a superconducting electromagnet. A magnetic field is produced when electrical current flows through a coil of wire. To get a large magnetic field, it is necessary to have a great deal of current flow through a lot of wire that comprises the coil. Before superconducting magnets existed, magnetic fields were limited. The wire would get very hot, and cooling was a great problem. Now the wire is made from a superconducting metal, such as niobium. Once the flow of electrons around the coil is started, the two ends of the coil are joined. The electrons keep whizzing around coil. Because there is no resistance, there is no dissipation of energy, and no additional electricity needs to be added to the coil. Without superconductivity, there would be no MRI.

One great hope is to make electrical transmission lines that are superconducting. Such transmission lines would eliminate electrical loss in power. It would be possible to move electricity over much greater distances than is possible today. The problem is that the metallic superconductors need to be so cold that they are not practical for transmission lines. There are new high temperature superconducting materials. They were discovered in 1986 by Karl Müller (1927–) and Johannes Bednorz (1950–). They were awarded the Nobel Prize in Physics in 1987 "for their important break-through in the discovery of superconductivity in ceramic materials." To date, superconductivity in these ceramic materials is not understood theoretically. Such materials can be superconducting at temperatures as high as 138° K. This temperature is high enough to make many applications practical. Because the high temperature superconductors are ceramic, the materials cannot be formed into wire as metals can. However, future advances may produce usable high temperature superconductivity, which will revolutionize power transmission and other fields of electronics.

20

Think Quantum

WHEN A PARENT HOLDS A TODDLER and points to the moon saying "that's the moon," the toddler certainly has an awareness of the bright object in the sky. The toddler may learn that the light in the sky is called "moon," but has no understanding of what and where the moon is. By the time a child is seven or eight, understanding of the moon has grown considerably. The child knows that it is very different from the street light at the end of the block, that it is very far away, and that you cannot touch the moon or even go there, although many years ago people actually did go to the moon. As an adult, a person has a good understanding of what the moon is even without knowing how to use Newton's classical mechanics to calculate its orbit around the Earth. The adult knows that the moon's apparent motion across the sky is caused by the rotation of the Earth, that the moon is far away, but not nearly as far as the planets in the solar system, and that a person feels much lighter and can jump higher on the moon than on Earth because the moon has less mass and, therefore, it exerts a weaker gravitational pull.

EXPERIENCE TEACHES US TO
UNDERSTAND THE CLASSICAL WORLD

As we grow up, our increased understanding of the moon comes
not just from education, but from the intuitive logic in the description of the moon in orbit around the Earth. This description is, in
many respects, consistent with everyday experiences. If you throw a
baseball, it arcs before falling back to Earth. If you throw the ball
harder and aim higher, it goes higher and farther, making a bigger
arc before hitting the ground. It is a simple and reasonable exten-
sion to accept that if you use a rocket and get an object moving
really fast, aiming very high, you can make the arc extend halfway
around the Earth, which is the basis for intercontinental missiles.
After that, it is not a great leap to accept that if you use an even
bigger rocket and get an object going even faster, the object will
have an arc that is more or less a circle going around the Earth in
orbit. Then the moon is just a very large object moving very fast so
that it is in orbit around the Earth.

The fact that we can progress from a baseball to the moon orbiting the Earth is based on everyday classical mechanics experiences.
However, it does require abstract reasoning to put the pieces together. The ancients reasonably thought that the moon circled the
Earth. After all, you can see it move across the sky. We can do a
simple experiment to see why the moon appears to circle the Earth.
Standing in the middle of a room with a single light on a wall, if
you slowly spin around, you will see the light come and go. With
your back to the light, you do not see it. As you rotate, the light will
come into the edge of your vision, move to the center of your view,
and then disappear at the other edge of your view. It does not appear
again until you have made a half circle of your rotation. Given this
simple experience combined with what we know about baseballs
and intercontinental missiles, it is easy to accept and, in fact, understand that the moon is in orbit around the Earth and that the Earth
rotates on its axis causing the "rising" and "setting" of the moon.

Our experiences and the basic nature of systems that obey classical mechanics allow us to develop an intuition for the behavior of many things we see about us daily. Even a novice pool player quickly grasps that if you hit the cue ball so that it strikes the left side of another ball, that ball will take off to the right. The pool ball collision is classical, and the balls move according to the precepts of classical mechanics with well-defined trajectories. However, the world around us, which is governed by the rules of quantum mechanics, is generally beyond consideration and understanding. When it comes to phenomena that are controlled by the properties of absolutely small systems, most people are like the toddler looking at the moon. They see it, but lack understanding of what they are seeing.

UNDERSTANDING WHAT WE SEE AROUND US REQUIRES SOME KNOWLEDGE OF QUANTUM MECHANICS

Why should we care? It is possible to go through life and see the moon with no idea of what it really is. A person can get up in the morning, go to work, eat, sleep, have a family, not know what the moon is, and be perfectly happy. It is also possible to have no conception of what makes the everyday things around us have their properties. We live in a sea of physical phenomena that can toss us around like ocean waves. We may not be able to control the physical world around us, but is it reasonable to completely lack any understanding of it? Do we want to be like a toddler or, even worse, an adult with no understanding of the moon? Do we really want to have no conception of why the element in an electric stove gets hot? I believe that the world is a more interesting place when we have some appreciation for the nature of the matter that surrounds us. From biological molecules to electrical conductivity, the natural world is driven by quantum phenomena. If we are adrift in an ocean of quantum physics, then some knowledge of quantum theory increases our appreciation for the wonders of the natural world.

COLOR IS A QUANTUM PHENOMENON

After plowing through the previous chapters, you have grown from toddler to adult in your quantum thinking. Now you know what color is. Let's go back to the first sentence in the book. Why are cherries red and blueberries blue? The question is what gives objects color and what makes one thing have a different color than another. The answer is that matter is made of atoms and molecules. In contrast to classical mechanics, where energy varies continuously, atoms and molecules have discreet energy levels. Light is also not continuous. Light comes in discreet packets called photons. A photon has a particular energy, which means it has a particular color. Because energy must be conserved, a photon can only be absorbed by the atoms and molecules that make up matter if the *energy of the photon matches the energy difference between two atomic or molecular quantum energy levels.* If the energy matches, the photon can be absorbed, and the system makes a transition from a lower energy level to a higher energy level. The photons that do not match energy level differences reflect from the object. So if the energy level spacing of the molecules is such that red light is absorbed, blue light will bounce off, and the object looks blue. If the energy level differences are such that blue light is absorbed, then red light bounces off, and the object looks red.

ENERGY LEVELS AND COLOR COME
FROM THE WAVELIKE NATURE OF PARTICLES

To reprise the color of objects in a little more detail, we discussed the one-dimensional particle in a box problem in Chapter 8. We learned that absolutely small "particles" are not particles in the everyday classical sense. They are actually waves or wave packets that are more or less localized in space. In the particle in a box, only waves of certain shapes were allowed. In three-dimensional systems, such as the hydrogen atom discussed in Chapter 10, the shapes of the waves are more complicated, but again, only certain shapes, called orbitals, can exist. This is also true for larger atoms and molecules, where the molecular electron waves were described as molecular orbitals. The electron waves (wavefunctions) in an atom or molecule have associated with them well-defined energies, the energy levels. We say energy is quantized; the energy changes in discreet steps. The discreet quantum energy levels are one of the major departures from classical mechanics. In classical mechanics, energy changes continuously.

We solved the quantum particle in a box problem and found that the energy levels depend on the size of the box. A larger box (larger molecule) has energy levels that are more closely spaced than a smaller box. The result, which applies to real molecules as well as the particle in a box, is that large molecules tend to absorb light in the red part of the spectrum. Red light is low energy light, and for large molecules there are relatively small energy differences between the energy levels. Smaller molecules absorb light in the blue because the molecular energy level differences are larger, and blue light is higher energy than red light. Really small molecules, like benzene (see Chapter 18), absorb in the ultraviolet part of the spectrum. Thus, they do not absorb visible light. Crystals of small molecules like naphthalene (mothballs) look white because none of the visible wavelengths can be absorbed. The energy levels are too widely spaced to absorb visible light. All of the visible light bounces off of the crystals, and they look white. This is the reason that salt crystals in a salt shaker are white. It is also the reason sugar crystals are white. Both salt and sugar have widely spaced energy levels that absorb in the ultraviolet, and reflect all visible colors of light.

QUANTUM PHENOMENA HOLD ATOMS
TOGETHER TO MAKE MOLECULES
AND DETERMINE THEIR SHAPES

We now know what holds atoms together to make molecules, what gives molecules their shapes, and why details of the shapes are important. We have seen that the electron waves of atoms combine to make molecular orbitals. Sharing the electrons among atoms in molecular orbitals can result in chemical bonds that hold atoms together to form molecules. In Chapters 12 through 14, we looked at molecular orbitals in some detail. We learned that molecular orbitals come in two flavors, bonding and antibonding. Appropriately placing electrons in the simple molecular orbital energy level diagrams can provide a great deal of information. In the hydrogen molecule (Chapter 12), the two electrons from the two hydrogen atoms go into the lowest energy molecular orbital, the bonding MO. The result is a covalent bond in which the atoms share a pair of electrons. However, the same considerations enabled us to see why a diatomic helium molecule does not exist. Each helium atom contributes two electrons to the hypothetical helium diatomic molecule. The first two go into the bonding MO but because of the Pauli Exclusion Principle, the next two electrons must go into the antibonding MO. The result is no net bonding, and the He_2 molecule does not exist. The covalent chemical bond is intrinsically a quantum phenomenon with no classical mechanics explanation.

For atoms larger than hydrogen, combining different s and p atomic orbitals produced hybrid orbitals with various shapes. The combination of different-shaped hybrid atomic orbitals to form molecular orbitals is responsible for the types of bonds that are formed (single, double, triple) and the shapes of molecules. We paid particular attention to organic molecules, which are molecules formed mainly from carbon, hydrogen, oxygen, and several other atoms. Organic molecules are important because they are the basis for life, as well as for materials such as plastics. We found that the types of bonds are very important. A molecule can readily rotate around a carbon-carbon single bond, changing the molecule's shape, but a molecule cannot rotate around a carbon-carbon double bond. The inability of organic molecules to rotate around carbon-carbon double bonds is crucial in biology.

In Chapter 16, we discussed fats. Double bonds made all the difference. Fats with double bonds cannot change their shape around the double bond. Polyunsaturated fats have multiple double bonds. All naturally occurring fats, except those from ruminants, have cis double bonds. That means the fat molecule is bent at the double bond. However, chemical processing of polyunsaturated fats to produce monounsaturated fats generates trans double bonds. Fats with trans double bonds are called trans fats. The trans fat molecule is straight at a trans double bond rather than bent. This difference in shape, which is produced by the properties of the quantum mechanical covalent double bond, has a substantial influence on the biological activity of the molecules. Trans fats have been linked to a variety of deleterious effects on human health. The shapes of biomolecules, such as proteins, are central to biology. The shapes of molecules are controlled by the quantum mechanical interactions between atoms that give rise to different types of molecular orbitals and different types of bonds. Therefore, the processes of life are controlled by quantum mechanics.

CARBON DIOXIDE IS A GREENHOUSE GAS BECAUSE OF QUANTUM EFFECTS

We have seen that the greenhouse effect produced by carbon dioxide, which is driving global climate change, is fundamentally quantum mechanical in nature. Carbon dioxide is a perfect storm of quantum effects that produces a vicious greenhouse gas. Hot objects give off light that we call black body radiation. The colors given off by a hot object could not be explained with classical theory. In fact, the classical theory was so far off that it was called the Ultraviolet Catastrophe because the theory predicted that an infinite amount of energy would be given off in the ultraviolet portion of the spectrum by any hot object. Clearly a hot object does not and cannot give off an infinite amount of energy. So the failing of classical theory was monumental. In 1900 Planck made the first use of the idea of quantized energy levels for electrons in matter to explain black body radiation. He was able to derive a formula for the colors given off by a hot object that matched experiments virtually perfectly. The hotter an object is, the more high-energy photons it emits. However, Planck's quantum theory showed that the amount of energy is not infinite and told us precisely how much light is emitted at each color. Stars are very hot, so they give off light in the visible and ultraviolet portions of the spectrum. An example of the black body spectrum for our Sun is shown in Figure 9.1. The Sun is a medium hot star. It appears slightly yellow. Very hot stars are blue and cooler stars than our Sun are red.

The Earth also emits black body radiation, but because it is very cool compared to a star, we cannot see the light the Earth emits with our eyes. The Earth's black body spectrum is shown in Figure 17.1. The light emitted by the Earth is in the infrared, that is, the long wavelength (the low energy) part of the spectrum. Without the atmosphere, all of the black body radiation emitted by the Earth would go into space, and our world would be much colder, perhaps too cold for human life. However, the atmosphere absorbs some of the black body radiation, trapping the heat, and keeping the Earth warm. Most of the heat trapping is done by water, which has quantized rotational energy levels that can undergo transitions in the very far infrared (long wavelengths and low energy). We have not mentioned quantized rotations before. Here is where your quantum intuition comes into play. We have talked about quantized electronic energy levels and quantized vibrational energy levels. Classical objects can rotate, for example, a top. The classical energy associated with rotation is continuous. Spin the top a little bit faster, and the energy is increased a little bit. It should come as no surprise that molecules in the gas phase, such as water vapor in the air, can rotate, and because a water molecule is absolutely small, the rotational energy is quantized. Rotational energy can only change in discreet steps. A water molecule can rotate at one speed and then have a step to another speed, but it cannot rotate at speeds in between. Think of what this would mean if it applied to big classical systems. You are riding you bike. You can pedal at one speed, but you can't pedal just a little bit faster. You would have to take a discreet step to the next quantized rotational energy level. Of course this doesn't happen for absolutely large objects for which energy is continuous.

Water does not absorb the Earth's black body radiation at the peak of the black body spectrum where a lot of energy is emitted. However, carbon dioxide does. As we discussed in Chapter 17, molecules have quantized vibrational energy levels. Carbon dioxide, CO_2, is made up of three atoms, with the carbon in the center. It is a linear molecule that can undergo bending vibrations. The vibrational motion has quantized energy levels. By happenstance, the energy difference between two CO_2 bending vibrational energy levels falls at the energy of the peak of the Earth's black body light emission. Therefore, the CO_2 molecules in the air absorb a significant part of the Earth's radiated black body energy that would otherwise go into space. The more CO_2 in the air, the less radiated energy escapes the Earth's atmosphere. The result is that as the amount of CO_2 in the air increases, more and more of the Earth's heat is trapped in the atmosphere, and the planet warms. CO_2 is a greenhouse gas because of the two quantum phenomena, black body radiation and quantized vibrational energy levels.

VERY HOT OBJECTS GIVE OFF
VISIBLE BLACK BODY RADIATION

While we are on black body radiation, now you know that whenever you see something glowing red, like molten lava coming out of a volcano or the hot element in an electric stove, you are seeing black body radiation. When you turn an electric stove on low, the temperature is low enough that all of the black body radiation is emitted in the infrared, and you can't see it with your eyes. If you used a spectrometer and an infrared detector, you could measure the IR colors emitted. The spectrum of the IR black body radiation from the stove element would tell you the temperature. When you turn the stove to high, the element turns red because it is much hotter. Most of the black body radiation is still in the IR, but the high-energy portion of the black body spectrum is in the low-energy portion of the visible spectrum, that is, it is red.

ELECTRICAL HEATING
IS A QUANTUM PHENOMENON

But why does the stove element get hot at all when electricity is passed through it? In spite of the fact that the stove element itself is a large object, we saw in Chapter 19 that electrical conductivity and electrical resistance, which produces the heat, are consequences of fundamental quantum effects. Metal crystals, such as sodium or copper, have electrons in atomic orbitals that interact with each other. The atomic orbitals from the atoms in the entire crystal combine to make molecular orbitals that spatially span the crystal. Like the aromatic molecule benzene that has six electrons in six delocalized molecular orbitals formed from interacting carbon p orbitals (Chapter 18), the electrons are not associated with a particular atom or pair of atoms. Rather, the MOs span the system, and the electrons are free to roam about the entire system, that is, a benzene molecule or a metal crystal. For benzene, six interacting atomic orbitals give rise to six molecular orbitals that are delocalized over the molecule. In benzene, with only six MOs, the energy spacing between the MOs is large. In even a very small metal crystal, there are billions and billions of atoms that give rise to billions and billions of MOs. Because there are so many MOs, they are very closely spaced. In a metal, all of these MOs form a band of quantum energy states called the conduction band. Each of these MOs is spread over the entire crystal. However, we know that these energy quantum states, the energy eigenstates, can be superimposed to form electron wave packets that are more or less localized in a manner consistent with the Heisenberg Uncertainty Principle. These electron wave packets are virtually free to move about the crystal.

Electrons are negatively charged. When a battery or other electrical source is connected to a piece of metal, for example a piece of copper wire, the electrons will be drawn to the positive battery electrode and move away from the negative battery electrode. The electrons are accelerated toward the positive side of the battery, which increases their kinetic energy. However, the electrons are not the only types of wave packets moving about in a metallic crystal. The mechanical vibrations of atoms in a crystal lattice have quantized energy levels. As with the electron bands in a macroscopic piece of metal, because there are so many atoms, there are a vast number of quantized vibrational levels that form bands of mechanical energy levels. The quantized delocalized mechanical motions of the coupled atoms in a lattice are called phonons. These delocalized phonon waves combine to form phonon wave packets. These phonon wave packets propagate through the lattice.

Electron wave packets and phonon wave packets collide. The collision is called an electron-phonon scattering event, illustrated in Figure 19.7. Some of the extra kinetic energy that the electron picked up because of its acceleration by the electric field is transferred to the phonon. The electron has less energy and the phonon has more energy after the scattering event. Many such electron-phonon scattering events cause the bath of phonons to have increased energy. Mechanical energy is heat. Temperature is a measure of the amount of kinetic energy in a substance. So the electron-phonon scattering events slow the electrons, which is what we call electrical resistance. The increase in phonon energy increases the temperature of the metal; it gets hot. The heating of a piece of wire as electricity (electrons) flows through it is caused by the collisions of electron and phonon wave packets. The scattering of these two types of wave packets is an intrinsically quantum mechanical phenomenon. The more electricity (current) that flows through the metal, the more collisions occur, and the hotter the metal gets. This is what happens when you turn up an electric stove. You are increasing the current (number of electrons flowing), and therefore, you are increasing the number of electron-phonon scattering events. Increasing the number of electron-phonon scattering events increases the energy transformed into heat, which causes the temperature to rise. When the metal stove element gets hot enough, it glows read because the black body emission has moved into the visible part of the spectrum. The net result is that turning on an electric stove or electric space heater, and seeing the heating element emit red light, involves a number of quantum phenomena. So every time you see a red hot stove element, rather than it being a complete mystery like the moon to a toddler, think quantized electron states, electron wave packets, phonon wave packets, electron-phonon scattering generating heat, and finally black body radiation. Every day observations are filled with quantum phenomena.

ABSOLUTELY SMALL

All of the quantum physical phenomena that surround us come ultimately from the fact that size is absolute, and absolutely small particles just don't behave the way classical objects, that is, absolutely big objects, behave. Baseballs are classical particles. Sound waves are classical waves. Baseballs and sound waves are big. In classical mechanics, the theory of big things, we have waves and particles. We discussed that light comes in discreet packets called photons. The description of photons and electrons as wave packets is profoundly different from anything in classical mechanics. Absolutely small particles, such as photons and electrons, are neither waves nor particles, as we saw in Chapters 4 through 7. They are wave packets. Sometimes they behave as waves (light diffraction from a grating and electron diffraction from a crystal surface) and sometimes they behave as particles (photons in the photoelectric effect and electrons in the cathode ray tube of old-style televisions). In fact, the essence of the nature of absolutely small particles is that they are really neither particles nor waves but a strange type of entity that has simultaneously particle and wave properties. This duality of matter is contained in the Heisenberg Uncertainty Principle. In contrast to a classical object like a baseball, we cannot know exactly simultaneously the position and the momentum (mass times velocity) of an electron or other absolutely small particles.

When is a particle absolutely small and subject to the new world of quantum physics? Dirac taught us that there is a minimum disturbance that accompanies a measurement, a disturbance that is inherent in the nature of things and can never be overcome by improved experimental technique. If the disturbance is negligible, then the object is large in an absolute sense, and it can be described by class physics. However, if the minimum disturbance accompanying a measurement is nonnegligible, then the object is absolutely small, and its properties fall in the realm of quantum mechanics. The quantum properties of absolutely small particles are not strange; they are just unfamiliar and not subject to our classical intuition. They are like the moon to the toddler. In this book the fundamental concepts of quantum theory have been explicated and applied to some important everyday phenomena. You are no longer a quantum toddler.

Glossary

absolute size—An object is large or small not by comparison to another object, but rather by comparison to the intrinsic minimum disturbance that accompanies a measurement. If the disturbance is negligible, the object is large in an absolute sense. If the intrinsic minimum disturbance is nonnegligible, the object is absolutely small.

absorption of light—The process by which the amount light is reduced, and the energy of an object is increased. Light (photons, particles of light) of the proper frequency (color) will cause the quantum state of an object to go to a higher energy state. The increase in energy of the object exactly matches the decrease in energy of the light. Absorption of light by objects is responsible for their color.

ångström—A unit of length that is 10^{-10} m (one ten billionth of a meter). The ångström unit has the symbol, Å.

anion—A negatively charged atom or molecule, such as Cl^-, the chloride anion. An anion is formed by adding one or more negatively charged electrons to a neutral atom or molecule.

atomic number—The number of protons (positively charged particles) in an atomic nucleus. A neutral atom (not an ion) has the same number of electrons (negatively charged particles) as protons.

atomic orbital—The name given a wavefunction (probability amplitude wave) that describes the probability distribution of an electron about an atomic nucleus.

black body radiation—The light given off by a hot object. The colors of the light depend on the temperature of the object. Black body radiation is the first physical phenomenon described with the ideas that became quantum mechanics by Max Planck in 1900.

Born interpretation—The description of quantum mechanical wavefunctions as probability amplitude waves. The Born interpretation, also referred to as the Copenhagen interpretation, states that the quantum mechanical wavefunctions obtained from solving the Schrödinger equation describe the probability of finding a particle in some region of space.

cation—a positively charged atom or molecule, such as Na^+, the sodium cation. A cation ion is formed by removing one or more negatively charged electrons from a neutral atom or molecule.

classical mechanics—The theory of matter and light that was developed before the advent of quantum mechanics. It treats size as relative and cannot describe absolutely small particles (electrons, photons, etc.). It is a powerful theory that works extremely well for the description of large objects such as airplanes, the trajectory of a rocket, or bridges.

classical waves—Waves, such as water waves and sound waves, that can be described with classical mechanics. Electromagnetic waves, which are classical mechanics' description of light, are also a type of classical wave. The classical description of light as waves works well for radio and other types of waves, but it cannot properly describe the particle nature of light (photons) responsible for such phenomena as the photoelectric effect.

closed shell configuration—An atom has the number of electrons associated with its nucleus that correspond to one of the noble gases, which comprise the right-hand column of the Periodic Table. A closed shell configuration is particularly stable. The noble gases, also called the inert gases, have the closed shell configuration, and are essentially chemically inert. An atom can obtain a closed shell configuration by gaining or losing electrons to become ions or by sharing electrons with another atom in a covalent bond.

constructive interference—Waves combine (add together) to make a new wave in a manner that increases the amplitude of the total wave. For waves with different wavelengths, constructive interference will occur only in some regions of space. The wave can be large in some region because of constructive interference but small elsewhere.

Coulomb interaction—The interaction between electrically charged particles that gets smaller as the distance increases. The interaction decreases in proportion to the inverse of the distance. The Coulomb interaction causes particles with opposite charges (positive and negative) to attract each other (an electron and a proton), and like charges to repel (two electrons or two protons).

covalent bond—A chemical bond that holds atoms together because the atoms share electrons.

de Broglie wavelength—The wavelength associated with a particle that has mass. All particles have de Broglie wavelengths. For large particles like baseballs, the de Broglie wavelength is so small that it is negligible. So large particles do not act like waves. For small particles (electrons, etc.), the wavelength is comparable to the size of the particle. Therefore, small particles can exhibit properties that are wavelike.

destructive interference—Waves combine (add together) to make a new wave in a manner that decreases the amplitude of the total wave. For waves with different wavelengths, destructive interference will occur only in some regions of space. The wave can be large in some regions because of constructive interference but small elsewhere because of destructive interference.

Dirac's assumption—A minimum disturbance accompanies any measurement. This disturbance is not an artifact of the experimental method, but is intrinsic to nature. No improvement in technique can eliminate it. If the minimum disturbance is negligible, a particle is large in an absolute sense. If the disturbance is nonnegligible, the particle is absolutely small. Dirac's assumption has been demonstrated by many experiments to be true and is central to quantum theory.

double bond—A chemical bond in which two pairs of electrons are shared between two atoms. A double bond is stronger and shorter than a single bond.

eigenstate—A pure state of a system associated with a perfectly defined value of an observable called an eigenvalue. A system in an energy eigenstate, such as a hydrogen atom, has a perfectly defined energy. The hydrogen atom has many different energy eigenstates, which have different energies (energy eigenvalues). A system in a momentum eigenstate has a perfectly defined value of the momentum. Each eigenstate has a wavefunction associated with it. Eigenstates are the fundamental states of quantum theory.

electromagnetic wave—A wave composed of electric and magnetic fields that oscillate at the same frequency and propagate at the speed of light. Electromagnetic waves are the classical mechanics description of light. The classical theory of electromagnetic waves is useful in describing many aspects of light and radio waves, but it cannot describe many phenomena, such as the photoelectric effect.

electron—A subatomic particle that has a negative charge. It is one of the basic constituents of atoms and molecules. Its negative charge is the same size but opposite in sign from the positive charge of a proton. An atom has the same number of electrons and protons, so it has no overall charge. Adding an electron to an atom makes an anion with one unit of negative charge. Removing an electron from an atom makes a cation with one unit of positive charge.

energy levels—In atoms, molecules, and other quantum absolutely small systems, energy is not continuous. Energy changes can only occur in discreet steps. Each distinct discreet energy is an energy level.

excited state—The state of an atom or molecule that has a higher energy than the minimum. An excited state is created when an atom or molecule that starts in its lowest energy state absorbs a photon of the right frequency to place the system in an energy level above the lowest energy, which is referred to as the ground state. Excited states can also be generated by heat and other mechanisms to put energy into an atom or molecule.

free particle—A particle that has no forces acting on it. A moving free particle will go in a straight line because no forces, such as gravity or air resistance, affect its motion.

frequency—The number of repetitions of a recurring event per unit of time. For a wave, the frequency is the number of wave peaks that pass by in a given time. For waves traveling at the same speed, a high frequency corresponds to a short wavelength and a low frequency corresponds to a long wavelength. The wavelength is the distance between peaks of a wave. For light waves, the frequency is the speed of light divided by the wavelength.

ground state—The lowest energy state of an atom or molecule. An excited state is created when an atom or molecule that starts in its lowest energy state absorbs a photon of the right frequency to place the system in an energy level above the lowest energy, the ground state. Excited states can also be generated by heat and other mechanisms to put energy into an atom or molecule.

Heisenberg Uncertainty Principle—The momentum and position of a particle cannot be known exactly simultaneously. If the momentum of a particle is known exactly, then the position is completely uncertain, that is, there can be no information on the position. If the position is known exactly, there can be no information on the magnitude of the momentum. In general, the principle states that the position and momentum can only be known within a certain degree of uncertainty. This is intrinsic to nature and not a consequence of experimental error.

hybrid atomic orbitals—Combinations (superpositions) of atomic orbitals that create new atomic orbitals with different shapes. Hybrid atomic orbitals are important in chemical bonding. Hybrid atomic orbitals will be formed to bond atoms together to produce a molecule with the lowest energy (most stable molecule). The shapes of the hybrid orbitals determine the shapes of molecules.

hydrocarbon—A molecule composed of only carbon and hydrogen, such as methane (natural gas) and oil.

inert gases (noble gases). Atoms such as helium, neon, argon, etc., that have closed electron shell configurations. They occupy the right-hand column of the Periodic Table of Elements. Because of closed shell configurations, they are essentially chemically inert. They do bond to other atoms to form molecules.

interference of waves—The combination of two or more waves to give a new wave. The waves can constructively interfere in some region of space to give an increased amplitude (larger wave) and destructively interfere in other regions of space to produce decreased or zero amplitude.

Joule—A unit of energy. One Joule (J) is a meter times kilograms squared divided by seconds squared. $J = m\ kg^2/s^2$.

kinetic energy—The energy associated with motion. A moving particle has kinetic energy equal to one half times the mass times the velocity squared, as in $E_{ke} = 1/2mV^2$.

light quanta—A single particle of light. A phonon.

lone pair—A pair of electrons in a molecule that occupies an atomic orbital but does not participate in bonding. Lone pair electrons are not shared between atoms.

molecular orbital—A wavefunction for a molecule composed of a combination of atomic orbitals (atomic wavefunctions) that span the molecule. Molecular orbitals can be bonding (bonding MO). Electrons in bonding MOs make the energy of the molecule lower. Molecular orbitals can be antibonding (antibonding MO). Electrons in antibonding MOs increase the energy of a molecule. To have a stable molecule, there must be more electrons in bonding MOs than in antibonding MOs.

momentum eigenstate—The state of a particle with perfectly defined momentum. A momentum eigenstate of a free particle, like a photon or electron, has a wavefunction that is delocalized over all space. The momentum can be known exactly but the position is completely uncertain. Momentum eigenstates can be superimposed (added together) to make a wave packet that has a more or less well-defined position.

nanometer—A unit of length that is one billionth of a meter, 10^{-9} m.

node—For a one-dimensional wave, a point where the amplitude of the wave is zero. For a three-dimensional wave, a node is a plane or other shaped surface where the wave amplitude is zero. The sign of the wavefunction changes when a node is crossed. In quantum mechanics, a node in a wavefunction describing a particle, such as an electron, is a place where the probability of finding the particle is zero.

optical transition—The change in state from one energy level to another in an atom or molecule caused by the absorption or emission of light.

orbital—Another name for the quantum mechanical wavefunction that describes an electron or pair of electrons in an atom or molecule. An atom has atomic orbitals, and a molecule has molecular orbitals.

particle in a box—A quantum mechanical problem in which a particle, such as an electron, is confined to a one-dimensional box with infinitely high and impenetrable walls. The energy levels of a particle in a box are quantized, that is, there are discreet energy levels. The particle in a box is the simplest quantum mechanical problem in which a particle is confined to a small region of space and has quantized energy levels.

Pauli Exclusion Principle—The principle that at most two electrons can be in an atomic or molecular orbital. If two electrons are in the same orbital, they must have opposite spins, that is, different electron quantum numbers s (one $+1/2$ and the other $-1/2$). The Pauli Exclusion Principle is important in determining the structure of the Periodic Table of Elements and the properties of atoms and molecules.

phase—The position along one cycle of a wave. The peak of the wave (point of maximum positive amplitude) is taken to have a phase of 0 degrees ($0°$), then the first node (point where the amplitude is zero) is $90°$. $90°$ is a quarter of a cycle of a wave. A phase of $180°$ is one half of a cycle. It is the point of maximum negative amplitude. Two waves of the same wavelength are said to be phase shifted if the peaks don't line up.

photoelectric effect—The effect explained by Einstein in which a single particle of light, a photon, can eject a single electron from a piece of metal. Einstein's explanation of the photoelectric effect showed that light is not a wave as described by classical electromagnetic theory.

photon—A particle of light.

Planck's constant—The fundamental constant of quantum theory. It is designated by the letter h. It appears in many of the mathematical equations found in quantum mechanics. For example, $E = h\nu$ says that the energy is the frequency ν (Greek letter nu) multiplied by Planck's constant. Planck's constant has the value h = 6.6×10^{-34} J s (Joule times seconds). Planck introduced the constant in 1900 in his explanation of black body radiation.

potential energy well—A region in space where energy is lowered because of some type of attractive interaction. A hole in the ground is a gravitational potential energy well. A ball will fall to the bottom, lowering its gravitational energy. It will require energy to lift the ball out of the hole. Electrons are held in atoms by a Coulomb potential energy well, that is, by the electrical attraction of negatively charged electrons for the positively charged nucleus. It requires the addition of energy to remove an electron from an atom. Enough energy can raise the electron out of the Coulomb potential energy well created by the attraction to the positively charged nucleus.

probability amplitude wave—Quantum mechanical wave (wavefunction) that describes the probability of finding a particle in some region of space. A probability amplitude wave can go positive and negative. The probability of finding a particle in some region of space is related to the square (actually the absolute value squared) of the probability amplitude wave. The greater the probability in some region of space, the more likely the particle will be found there.

proton—A subatomic particle that has a positive charge. It is one of the basic constituents of atoms and molecules. Its positive charge is the same size but opposite in sign from the negative charge of an electron. An atom has the same number of electrons and protons, so it has no overall charge. The number of protons in an atomic nucleus, called the atomic number, determines the charge of the nucleus. Different atoms have different numbers of protons in their nuclei.

quantized energy levels—Energy levels that come in discreet steps. The energy is not continuous. Atoms and molecules have quantized energy levels.

quantum number—A number that defines the state of a quantum mechanical system. There can be more than one quantum number to fully specify the state. In an atom, each electron has four quantum numbers, n, l, m, and s, which can only take on certain values. Quantum numbers arise from the mathematical description of quantum mechanical systems.

radial distribution function—The mathematical function that describes the probability of finding an electron a certain distance from the nucleus of an atom independent of the direction. It is obtained from the wavefunction for the electron in the atom.

Rydberg formula—The early empirical formula that gave the colors of light emitted or absorbed by hydrogen atoms.

Schrödinger Equation—A fundamental equation of quantum theory. Solution of the Schrödinger Equation for an atom or molecule gives the quantized energy levels and the wavefunctions that describe the probability amplitude of finding electrons in regions of space in an atom or molecule.

single bond—A chemical bond that holds two atoms together through the sharing of one pair of electrons.

size, absolute—An object is large or small depending on whether the intrinsic minimum disturbance that accompanies a measurement is negligible or nonnegligible. If the minimum disturbance is negligible, the object is large in an absolute sense. If it is nonnegligible, it is absolutely small. Absolutely small objects can be described by quantum mechanics, but not by classical mechanics.

size, relative—Size is determined by comparing one object to another. An object is big or small only in relation to another object. In classical mechanics it is assumed that size is relative. Classical mechanics cannot describe objects that are small in an absolute sense.

spatial probability distribution—The probability of finding a particle, such as an electron, in different regions of space. The spatial probability distribution can be calculated from the quantum mechanical wavefunction for a particle.

spectroscopy—The experimental measurement of the amount and colors of light that are absorbed or emitted by a system of atoms or molecules.

Superposition Principle—"Whenever a system is in one state, it can always be considered to be partly in each of two or more states."

This quantum mechanical principle says that a system in a particular quantum state can also be described by the superposition (addition) of two or more other states. In practice, this generally means that a particular wavefunction can be expressed as the sum of two or more other wavefunctions. For example, the wavefunctions for molecules can be formed by the superposition of atomic wavefunctions. A photon wave packet can be formed by the superposition of momentum eigenstates

triple bond—A chemical bond that holds two atoms together by sharing three pairs of electrons. A triple bond is shorter and stronger (harder to pull the atoms apart) than a double or a single bond.

Uncertainty Principle—The statement that the momentum and position of a particle cannot be known exactly simultaneously. If the momentum of a particle is known exactly, then the position is completely uncertain, that is, there can be no information on the position. If the position is known exactly, there can be no information on the magnitude of the momentum. In general, the principle states that the position and momentum can only be known within a certain degree of uncertainty. This is intrinsic to nature and not a consequence of experimental error.

vector—A directed line segment usually represented by an arrow. A vector is a quantity with both magnitude and direction. A car going 60 miles per hour has a speed, which is not a vector. A car going 60 miles per hour north has a velocity, which is a vector because it has a magnitude (60 miles per hour) and a direction (north).

velocity—A vector describing both the speed and the direction in which an object is moving.

wave packet—A superposition of waves that combine to make a particle more or less located in a region of space. The superposition of waves has regions of constructive and destructive interference. The probability of finding the particle is large where there is constructive interference. The superposition of waves more or less localizes a particle in some region of space. The location cannot be perfectly defined because of the Heisenberg Uncertainty Principle.

wavefunction—A solution to the Schrödinger Equation for a particular state of a system, such as an atom or molecule. A wavefunction is a probability amplitude wave. It gives information on finding a particle in a particular region of space. For example, the wavefunctions for the hydrogen atom give the probabilities of finding the electron at different distances and directions from the nucleus.

wavefunction collapse—A state of a system is frequently a superposition of wavefunctions. Each wavefunction has associated with it a definite value of an observable, for example, the energy. Because a superposition is composed of many wavefunctions, it has associated with it many values of an observable. When a measurement is made, the system goes from being in a superposition of wavefunction to being in a single wavefunction with one value of the observable (e.g., the energy). It is said that the measurement causes the wavefunction to collapse from a superposition of states into a single state with one value of the observable. It is not possible to say beforehand which state the superposition will collapse into. Therefore, it is not possible to say ahead of time which value of the observable will be measured.

wavelength—The repeat distance in a wave, that is, the distance from one peak in the wave to another.

Index